W0018357

Creativity in Military Complexity

This work offers a groundbreaking exploration of the urgent need for creativity and innovation in contemporary military thought.

In an era characterised by the ceaseless flux of global dynamics, traditional paradigms of warfare have become increasingly obsolete. The pursuit of victory no longer lies in the fixation upon past conflicts but rather in the discerning assessment of and adaptation to the challenges that will shape the conflicts of tomorrow. This innovative work dissects the barriers that have thus far hindered the realisation of this potential. Furthermore, it challenges the status quo and advances a series of recommendations poised to steer international militaries towards success in the theatre of contemporary conflicts. Drawing from their extensive involvement with defence forces worldwide, the authors introduce concepts such as military design thinking as catalytic instruments of change. Through introspective reflections and real-world case studies, they present implications for mitigating cognitive biases, heralding a transformative epoch in military operations. It is this very transformation that furnishes militaries with the capacity to surge ahead of their adversaries, a capability proving to be indispensable in modern warfare.

Offering a well-illuminated path for military entities to adapt and flourish within an ever-evolving global landscape, this book will be of much interest to students of military studies, security studies, and international relations, as well as military professionals and leaders in the field.

Cara Wrigley is a Professor of Design within the Faculty of Engineering, Architecture and Information Technology at The University of Queensland, Australia. She has published a number of books, including *Design Thinking Pedagogy* (2022), *Design Innovation and Integration* (2021), *Design Innovation for Health and Medicine* (2020), and *Affected: Emotionally Engaging Customers in the Digital Age* (2018).

Murray Simons is the Chief of Air Force Fellow at the University of New South Wales, where he teaches and supervises military students at the Australian Air and Space Power Centre. He has a doctorate in professional military education.

Routledge Studies in Conflict, Security and Technology

Series Editors:

Mark Lacy
Lancaster University

Dan Prince
Lancaster University

Sean Lawson
University of Utah

The *Routledge Studies in Conflict, Technology and Security* series aims to publish challenging studies that map the terrain of technology and security from a range of disciplinary perspectives, offering critical perspectives on the issues that concern publics, business and policymakers in a time of rapid and disruptive technological change.

Fifth Generation Warfare
Dominating the Human Domain
Armin Krishnan

Technology and Governance Beyond the State
The Rule of Non-Law
Edited by Nicole Stremlau and Clara Voyvodic Casabó

Military Design Thinking
An Historical and Paradigmatic Analysis
Aaron P. Jackson

Theorising Cyber (In)Security
Information, Materiality, and Entropic Security
Noran Shafik Fouad

Creativity in Military Complexity
Design, Disruptors and Defence Forces
Cara Wrigley and Murray Simons

Digital (Dis)Information Operations
Fooling the Five Eyes
Edited by Melissa-Ellen Dowling

For more information about this series, please visit: www.routledge.com/Routledge-Studies-in-Conflict-Security-and-Technology/book-series/CST

Creativity in Military Complexity

Design, Disruptors and Defence Forces

Cara Wrigley and Murray Simons

Routledge
Taylor & Francis Group
LONDON AND NEW YORK

Designed cover image: Designed by Leon Fitzpatrick

First published 2025
by Routledge
4 Park Square, Milton Park, Abingdon, Oxon OX14 4RN

and by Routledge
605 Third Avenue, New York, NY 10158

Routledge is an imprint of the Taylor & Francis Group, an informa business

© 2025 Cara Wrigley and Murray Simons

The right of Cara Wrigley and Murray Simons to be identified as authors
of this work has been asserted in accordance with sections 77 and 78 of the
Copyright, Designs and Patents Act 1988.

The Open Access version of this book, available at www.taylorfrancis.
com, has been made available under a Creative Commons Attribution-Non
Commercial-No Derivatives (CC-BY-NC-ND) 4.0 license.

Any third party material in this book is not included in the OA Creative
Commons license, unless indicated otherwise in a credit line to the material.
Please direct any permissions enquiries to the original rightsholder.

Trademark notice: Product or corporate names may be trademarks or
registered trademarks, and are used only for identification and explanation
without intent to infringe.

British Library Cataloguing-in-Publication Data
A catalogue record for this book is available from the British Library

ISBN: 978-1-032-81949-5 (hbk)
ISBN: 978-1-032-81951-8 (pbk)
ISBN: 978-1-003-50218-0 (ebk)

DOI: 10.4324/9781003502180

Typeset in Times New Roman
by Apex CoVantage, LLC

Contents

Figures

Tables

Preface

In combat . . . if you're predicable, you're dead!

The contemporary geopolitical environment is a complex web of shifting interactions that cannot be resolved through traditional linear mechanistic approaches. Complex interactions between national interests, statecraft, and geopolitics mean that modern nations may be simultaneously engaged across the spectrum of competition at multiple points. No matter how a nation terms this new paradigm, it boils down to a more dynamic operating environment, which sees state and non-state actors employing instruments of power with novel, adaptive, and surprising efficiencies. To navigate this complex landscape, the only constant is change itself.

Various threat assessments underscore the importance of recognising change as a constant factor in this ever-evolving environment. One such threat is the internal necessity of fostering a strong sense of self-confidence. Although this attitude is essential in combat, it can sometimes be counterproductive when it comes to inter-agency collaboration and the construction of cohesive national power. A belief in the superiority of one's own methods can hinder openness to new ideas and impede effective cooperation.

Attitudes of military exceptionalism can give a false belief that change is rarely necessary. Wars have a way of testing our true effectiveness; however, in peacetime, the military often "marks its own homework" based on a self-generated rubric of success, with the result being a self-aggrandised opinion of its own performance, which undermines the need to promote innovation mindsets (an innovative mindset is a way of thinking that embraces creativity, openness to new ideas, and a willingness to take risks via experimentation).

Internal cultural change is an area where militaries have historically struggled. The emphasis on strict discipline and adherence to orders creates a self-stabilising influence that discourages open-mindedness. While an honest evaluation might be uncomfortable, it is essential when defending the nation's strategic interests. Therefore, there is a critical need to be more open to alternative perspectives.

Militaries cannot become complacent about the importance of antebellum (pre-war) innovation, especially against rising peer or near-peer adversaries. The West, though active in several asymmetric wars in recent years, now faces a looming multi-front threat of significant proportions. While technological advances continue, the crucial importance of cultural shifts remains underemphasised. In peacetime, the lack of urgency hinders transformative initiatives, making it difficult to rally efforts for considerable improvements, but changing this culture is essential, as transitioning to wartime innovation mindsets faces significant institutional inertia that must be addressed urgently.

The urgency of the present moment cannot be overstated; we can no longer afford to squander time. We should have been preparing for the impending challenges rather than busying ourselves with less important tasks. Just as many disruptors have revolutionised industries (e.g., Uber's impact on the taxi industry and Instagram's effect on Kodak), we must adopt a proactive approach to change. Our path forward necessitates embracing change, pivoting, and continuously learning. This is a crucial lesson we cannot afford to neglect if we wish to avoid being on the wrong side of history, and it begins now, within the context of this book.

This book is written for a diverse readership. It will interest those who recognise the need to enhance creativity and innovation within the military, but perhaps more importantly, it targets those who remain unconvinced. Many people resist change and prefer the predictability of their established systems, but our adversaries thrive on knowing how we will react. To maintain a competitive edge, we must remain dynamic and adaptable. Without innovation, militaries will inevitably soon *suffer what they must*.

This book is not just for senior leaders; decision-makers at all levels play a crucial role in fostering a culture of continuous improvement in the military. Everyone in the military has a vested interest in ensuring success, as innovation is everyone's responsibility. Additionally, this book is valuable for those on the periphery of national military power, including other government agencies, academics, the defence industry, and members of the wider community.

The concepts and literature contained in this book are Western, educated, industrialised, rich, and democratic (WEIRD), which emphasises that much of the world's scholarly work is derived from these nations and, therefore, most relevant to other WEIRD environments. The book is also heavily influenced by the two authors' firsthand experiences in Australasia; however, the discussion is based on global literature and extensive engagement with the international military design community.

For more than a decade, Professor Wrigley has worked in both the corporate and defence sectors on numerous major projects, reviews, and initiatives. As an industrial designer, Professor Wrigley helps companies design devices, services, systems, and business models to maintain global competitiveness. Within military circles, Professor Wrigley has facilitated major Australian Army, Navy, and Air Force projects in dealing with tactical, operational and strategic design challenges. Throughout these years, she has seen the good, bad, and ugly sides of military innovation. Her contribution to this book, therefore, represents many years of working with and observing how the military culture of innovation rises and falls. Thus, her real strength comes from being an objective outsider who senses opportunities. In this contestability role, she can see things others often miss.

The two authors first met when they were both teaching a design workshop at the Australian Command and Staff College in 2016. The then Wing Commander Simons was a New Zealand exchange officer on the teaching faculty and had already developed a strong interest in encouraging mid-career officers to expand their cognitive potential. His pedigree in cognitive psychology and educational theory dates back to the early 1980s but has been complemented by various university degrees and 35 years in uniform, including operational tours to the Middle East and Timor Leste. At the time of writing this book, he was the Air and Space Power Centre's Chief of Air Force Fellow at the Australian Defence Force Academy. Needless to say, the opinions expressed by the authors do not confer endorsement by any of their affiliated organisations.

Sharing a passion for innovation and creativity has brought their professional paths together time and again. Often labelled as "two frustrated academics," they lament the missed opportunities to tap into the potential of serving personnel. This book aims to be a catalyst for those who see the urgent need to improve the culture of innovation. In essence, the authors act as voices for the many frustrated service members who have shared their stories and expressed their dismay. These pages tell their story.

This book is not intended as an explicit roadmap for change. There is no one-size-fits-all action plan that will suit every unit, service, or national military. Instead, it serves as an initial step, prompting reflection and highlighting major challenges. While some high-level recommendations are offered, these should be used only as prompts. Bespoke structures, changes, and strategies should still be researched, created, and designed to suit local needs.

To grow, we must change, but change can be cognitively confronting for many. This book, therefore, will not be without critics. The nature of what we propose is challenging, making it an uneasy read for some. Over the decades, facing criticism has become part of the design process. This may not be a popular stance or win us many friends, but it is the core purpose of this book. Achieving meaningful, positive change requires engaging in these uncomfortable conversations. This isn't about us – our reputation, our egos – it's about the larger mission. Not every idea will fit neatly into every context, but having the courage to question, challenge, reflect upon, and critically examine existing practices is crucial. Our sincere hope is that this dialogue will contribute to the greater good and help the liberal-democratic world navigate the troubling years ahead.

Cara + Murray

Acknowledgements

We extend our heartfelt appreciation to those dedicated individuals who exemplify service above self. Your unwavering commitment to challenging the status quo inspires us all. Through your sacrifices, resilience, and determination, you have significantly contributed to the ongoing evolution of our defence forces, shaping the discourse in these pages.

To those who generously shared their stories and insights, we offer our deepest gratitude. Your contributions have been instrumental in developing the ideas and perspectives presented in this book. We also sincerely appreciate the defence forces we studied throughout this work; your service ensures the freedoms and liberties that enable genuine discourse to occur.

We convey our sincere thanks to our readers for joining us on this journey. We hope the insights shared here will ignite meaningful conversations, challenge established paradigms, and inspire renewed creativity and innovation in military thinking. The individuals highlighted in this work (Figure 0.1) represent the many creative minds who have shaped, inspired, and encouraged us throughout the years. Their dedication serves as a constant reminder to never give up on the relentless pursuit of greater innovation.

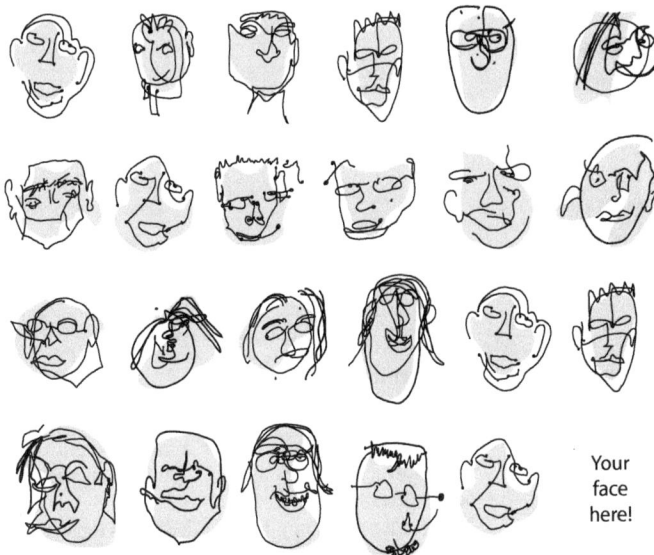

Your
face
here!

Figure 0.1 Faces of Thanks

Glossary

Terminology	Definition
capability advantage	Superior resources, technology, or training, enabling the successful objectives of a military entity to be met more effectively and efficiently than those of its adversaries.
cognitive bias	The brain's tendency to default to previous responses for a perceived similar situation, which is typically only recognised when the response is wrong.
complex	A highly interdependent system where impacting one part might have unpredictable and nonlinear impacts on other parts of the system.
complex adaptive system	A dynamic network of interconnected elements that constantly evolves and adapts in response to changes and interactions within its environment.
complicated	A system with multiple independent components that work in a linear and predicable way. The more components there are, the more complicated the system.
convergent thinking	A cognitive process that focuses on finding a single correct solution to a problem, typically through logical reasoning and deductive methods.
creativity	The ability to generate new ideas, concepts, or solutions that are original, valuable, and often unexpected.
Cynefin	References a conceptual framework developed by David Snowden that assists decision-makers in navigating complexity and chaos in an environment.
design	The process of planning, creating ideas, and implementing these ideas to improve the artificial or natural environment.
design thinking	An umbrella term used to describe many user-centred approaches to problem solve collaboratively.
disruptors	Disruptors are agents of change – be they individuals (e.g., mavericks), technologies, or innovations – that challenge conventional practices, industries, or markets, often catalysing substantial shifts or transformations.

divergent thinking	Unrestrained and imaginative ideas to expand established mental models.
doctrine	Formalised ways of conducting activities based on proven success.
emergent	A modified system or object that represents an incremental change from the previous version.
emerging operating environments	The constantly dynamic and evolving system within which military operations are conducted.
emerging threats	Emerging refers to incremental modifications to a known system or object. It differs from novel threats, which define something completely new. An emerging threat is an increasing yet anticipated risk.
epistemic diversity	The inclusion and integration of different types of knowledge, perspectives, and cognitive approaches within a team or community to enhance problem-solving and innovation.
experiment	To test and adjust potential options to ensure they can succeed.
grey zone	Military behaviour below the threshold of *acts of war* but intended to push out normative tolerance levels.
ideation	The phase of design thinking where novel ideas are generated to help explore alternative approaches.
innovation	The introduction of new ideas or processes that bring about change or improvement at the tactical, operational, or strategic level.
maverick	An altruistic person who constantly seeks better ways of doing things.
novel	A completely new concept or object that has not evolved from earlier versions.
OODA	John Boyd's observe, orient, decide, and act model.
RBIO	international order synonymous with rules-based global order.
reflexive	A mode of thought or action characterised by introspection, self-awareness, and the ability to critically examine one's own beliefs, biases, and assumptions.
statecraft	The art of managing a country and its position on the world stage.
transient advantage	Similar to competitive advantage over competition, but reinforces the point that in complex systems, such gaps are never permanent and need constant investment to maintain.
VUCAN	Stands for volatile, uncertain, complex, ambiguous, and novel.
warfighter	A generic term for any member of the armed forces, regardless of rank or uniform colour.
wicked systems	A complex adaptive system where decision-makers are part of the system and, therefore, their actions (or potential actions) influence the system's behaviour.

1 Introduction

Are we here to protect the nation, or the status quo?

1.1 Opening the Fusillade

Throughout history, militaries have celebrated their greatest successes by repeatedly recounting heroic deeds. In fact, so culturally entrenched is this emphasis on tradition that inductees cannot help but take pride in such achievements of former days. This is for good reason – promoting traditions and mythologising heroes are a critical part of acculturating honour and loyalty. It even helps with resilience during challenging times and guides members in their moral decision-making. However, a potential price of this emphasis on traditions is subtly promoting a culture of *walking backwards into the future*. In military classrooms, an extension of this phenomenon is, in fact, so well known it is referred to as "learning to fight the last war." Other axioms question whether military courses exist to teach *thinking* or *remembering*. Nonetheless, the study of historical military innovations serves as an encouragement to future generations.

While many modern militaries prioritise major capital investments in large and expensive technology systems and capabilities, this alone is insufficient. The true essence of innovation lies not only in acquiring exquisite equipment but also in cultivating a cadre of innovative and creative thinkers capable of employing these systems in novel and effective ways. Furthermore, the challenge of transitioning between different systems also has to be factored into this cocktail of competing needs. Platform generations spanning multiple capability requirements might not meet future needs, hindering optimised force structures.

Under the "fight tonight" mantra, the moment chooses us – not the other way around. We stand and fight with what we have got, not with the idealised future integrated system. Although most acquisition plans have a transition for introduction into service, the bigger question is whether this planning includes the necessary agile mindset shift required to outclass an adversary who has

DOI: 10.4324/9781003502180-1
This chapter has been made available under a CC-BY-NC-ND 4.0 license.

seized the initiative by choosing the time and place that gives them the greatest advantage.

Fostering a culture of innovation and creativity within military ranks is paramount to ensuring readiness and effectiveness in an ever-changing operational landscape. After all, there is no acquisition requirement that ensures we are cognitively creative in how we utilise the platform agaainst the enemy.

Recent regional conflicts reinforce the absolute necessity of innovation, yet it is widely accepted that the West is ceding the initiative in this line of effort. Despite looming adversaries giving the appearance of being calcified hierarchically, they are also astutely studying our weaknesses and investing heavily in gaining full-spectrum advantages. In fact, in many cases, the Western primacy of peacetime conformity still overshadows the importance of innovation. This is particularly true in the bureaucratically-strangled world of peacetime military preparations. Other terms, such as the "valley of death" and "frozen middle," further highlight how widespread and recognised the frustrations truly are. It seems like everyone is aware of this systemic problem, yet few are either willing or capable of fixing it.

On the surface most Western militaries do invest a fair amount of time and money into promoting creativity and innovation. Even during peacetime, there are some remarkable examples, particularly in the technological domain. (Murray & Millett, 1998). Of note is the plethora of external innovation agencies (e.g., defence industries, independent think tanks, dedicated innovation centres) as well as numerous internal counterparts (e.g., Defense Advanced Research Projects Agency, Defence Science and Technology Group, and single service innovation hubs, etc.). Sadly, however, such incremental and often isolated interventions fail to make any meaningful difference to the military's organisational culture. This is arguably because they focus more on technology than on changing cognitive paradigms and concepts.

This book goes beyond the ubiquitous civilian industry's thoughts on promoting creativity. While corporate ideas can still be valuable and should be explored with enthusiasm, the unique nature of the military demands caution when adopting a "copy-paste" approach. Defence forces exist to defend their nations in even the darkest hours. They are not structured to return a profit, nor should their practices blindly mirror commercial businesses that do. There is currently a significant lack of dedicated research that tackle the big questions around promoting a military culture of innovation. In particular, there is a need to re-examine the pros and cons of many previously unquestioned military habits. Perhaps the greatest of these is the internal threat of unexamined self-stabilising influences that actively resist change.

Self-stabilising influences extend from tangible rules and consequences through to more subtle cultural norms that reinforce conformity. The requirement for doctrinal changes to be combat-proven underscores the difficulty of introducing new ideas. When military schools are limited to teaching doctrine that only reflects past successes, stagnation becomes inevitable, leaving no room to teach emerging leaders how to be creative or think outside the box. Thus, the concept of "learning to fight the last war" epitomises a hesitancy to embrace change during peacetime. This mindset regrettably fails to acknowledge that although the fundamental nature of warfare remains a constant, its character and intricacies have undergone a profound transformation in the 21st century (even the last few years). Wars are no longer confined to distant theatres, shielded from the purview of the global populace; instead, they are subject to intense international scrutiny, further magnified by the ubiquity of social media and real-time global communication networks.

For democratic nations, their mandate transcends merely succeeding in warfare but increasingly encompasses the vital and intertwined responsibility of conflict prevention. To excel in this multifaceted environment, Western military institutions must disentangle themselves from the confines of entrenched, convergent thinking. The imperative lies in either reconfiguring their conceptualisation of warfare or facilitating a transformation in the mindset of military strategists themselves. Such mindset changes are required across the three main fields of military innovation: peacetime, wartime, and technological (Rosen, 1991). While technological innovation is often outsourced, the real challenge for military personnel is bridging the gap between peacetime complacency and wartime urgency (Figure 1.1). "Train as you fight" must become more than just a slogan.

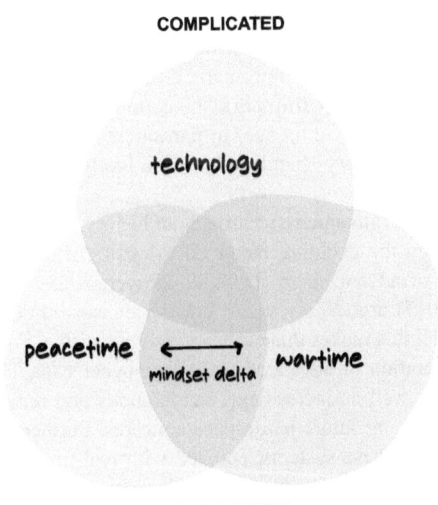

COMPLICATED

technology

peacetime ⟵⟶ wartime
mindset delta

COMPLEXITY

Figure 1.1 Delta of Innovation Mindsets

To navigate the complexities of contemporary conflict and flourish amidst the challenges posed by an ever-evolving landscape, Western militaries must earnestly embrace a paradigmatic shift in their problem-solving and decision-making processes.

Most nation-states face a number of traditional and non-traditional security challenges. While none of these are particularly new in the broad sense, their emergent level of sophistication demands increasingly innovative responses. The following six thematic clusters highlight how traditional issues are evolving rapidly:

1. **Changing geopolitical dynamics:** Threats to the rules-based international order (RBIO) have increased markedly in recent years, with great powers showing greater disdain for international conventions and treaties (Allison, 2020; Owen, 2021). This comes in stark contrast to the relative stability enjoyed during the 45 years of the Cold War (the period when RBIO was established) and then the following 30 years of comparatively minor regional conflicts. The current muscle flexing of revisionist powers, however, is raising the stakes of an imminent large-scale war. For example, geopolitical shifts and power realignments are altering the strategic landscape in the Indo-Pacific region (Kabutaulaka, 2010). This coincides with emerging environmental challenges that are hitting smaller regional countries particularly hard. Unsurprisingly, they are hedging their alliances to maximise economic support against this existential threat (Sora, 2022).

2. **Evolving security environment:** Both the global and regional security landscape are evolving rapidly with the jostling of great powers. In concert with these shifts in diplomatic alliances and economic interdependency, military competition is also heating up. The rise of countries seeking greater global influence, in particular, has seen a commensurate increase in military spending by these likely competitors (Augier et al., 2017).

3. **Technological advancements:** Throughout history, superior technologies have often been the deciding factor on the battlefield. Today, however, it is the rate of advancement (Dunk & Kruger, 2023) or, as Lawrence Freedman (2017) argues, the wider breadth of capabilities being developed in parallel that makes this race even more decisive. For example, the current proliferation of advanced weaponry, cyber capabilities, artificial intelligence, as well as increasingly autonomous and remote systems is significantly shaping future battlespace concepts. Furthermore, the interconnectedness of these systems, coupled with real-time intelligence and networked communications, is reducing decision times to faster than traditional human capabilities. Unsurprisingly, emerging competitors are also

investing heavily in such technologies, so militaries must seek alternative ways to achieve and maintain a competitive edge.

4. **Operational effectiveness:** The rapid advancement in technologies creates a threat to the operational effectiveness of traditional warfighting methods. For example, the conflict in Ukraine has seen significant impacts on armoured warfare doctrine as a result of ingenious Ukrainian drone modifications. Militaries must recognise the need to reimagine how they not only shape and deter but also respond to these emerging threats. Legacy thinking for future problems has an inevitable outcome. The West must become more innovative in updating its doctrine, concepts, and capabilities to align with the emerging complex and dynamic nature of conflict (Parker, 2020). The real challenge is simultaneously advancing both technology and ideas – after all, neither will produce an advantage in isolation.

5. **Complexity:** Traditionally, "wars of choice" have been fought on distant shores as a shaping activity – and thus out of sight from both politicians and the general population. The manner in which they were fought, whether on land or at sea, was typically through set-piece moves of closing and engaging with the enemy. The outcome of such attritional warfare was often predictable, based on numerical and weapon superiority. In short, the conduct of war was linear and reductionist. Commanders could set the conditions for success by picking the time and place of an engagement to suit their strengths. Today, however, wars are no longer fought solely from metaphorical trenches on a distant front line; rather, they are a complex mix of diplomacy, economics, alliances, and deep strikes well behind enemy lines. Since the advent of air power – and more recently space and cyber – winning wars is no longer about geographic lines on a map. Such industrial-era concepts have long since melted away, with all parties now avoiding the vulnerabilities of being merely complicated. Modern warfare requires military thinkers who not only survive, but also thrive in the realities of complex adaptive systems. Legacy mindsets of incrementalism are obsolete against these systems.

6. **Domestic pressures:** With automation progressively negating the need for transactional (complicated) jobs, industries are increasingly competing for society's most cognitively capable innovators (Dunk & Kruger, 2023). Meanwhile, automation in warfighting systems is translating into the militaries requiring smarter workforces. This squeeze on peacetime recruiting from a competitively diminished talent pool demands more innovative ways to attract and retain high-calibre personnel. Western militaries need to modernise not only the way they employ and incentivise people but also how they invest in developing their cognitive agility. Addressing national mobilisation demands a seismic step change again.

1.2   The Imperative for Change Amongst the Rise of Complexity

Militaries face the imperative of confronting the challenges presented by technological advancements and the asymmetric nature of warfare that has defined the past two decades. Yet the shaping of today's leaders by studying preceding conflicts, such as the near disaster of Korea and the contentious Vietnam War, underscores their pivotal role in sculpting the future force. Embracing innovation is paramount to ensuring military agility, adaptability, and readiness to confront the battles of tomorrow, thereby averting the recurrence of past mistakes and defeating future adversaries.

Faced with the need to adapt and respond to a rapidly evolving environment marked by intense and continual competition, this contestation brings forth a series of novel challenges that unfold amidst escalating uncertainty and complexity in national security. Importantly, these challenges can no longer be effectively addressed through traditional and functionalist approaches to strategy, force design, and capability development. The landscape has shifted. Today, militaries must embrace non-traditional methodologies to imaginatively formulate and test new theoretical frameworks that will enable them to generate and exploit advantages over adversaries in this emerging contest.

In this new contest, militaries must be adaptive, agile, and open to exploring unconventional means of enhancing their capabilities and maintaining a competitive edge. This entails fostering a culture of innovation, promoting cross-disciplinary collaboration, and leveraging emerging technologies to support non-traditional methods of strategy development and capability acquisition. By doing so, militaries can effectively navigate the complex and uncertain environment, ensuring they remain capable of meeting the evolving challenges and achieving their strategic objectives.

Complexity fundamentally changes the way militaries must think. For millennia, the profession of arms was comfortable in developing linear reductionist processes that worked perfectly for complicated set-piece battles where two sides lined up opposite each other. The past half-century, however, has seen the confluence of diplomacy, international law, domestic politics, and a myriad of invisible influences (e.g., cyber, grey zone, space power, among others). These entangled dimensions mean military commanders can no longer plan battles based purely on the mathematics of logistics or the geography of the battlefield. Nor can they effect cultural change simply by issuing orders from the comfort of their large office. Culture is complex and does not respond well to hierarchical commands from above. To positively influence a culture into becoming more innovative, complexity theory must be embraced.

Unlike complicated systems, where relationships are predictable and constant, the variables in a complex system are beyond even the best currently available artificial intelligence software. Understanding and thriving in complexity is the new archetype of the successful military commander. Not only must they exploit the apparent rules of complexity to generate cultural change, but they must also embrace the nature of complex systems if they are to be triumphant in defending the nation's strategic interests.

Complexity in the modern battlespace encompasses the intricate and multifaceted nature of military operations. It is characterised by a convergence of interconnected variables, diverse actors, and dynamic environments that present challenges and uncertainties. The modern battlespace is shaped by technological advancements, information warfare, hybrid threats, multi-domain operations, urban warfare, geopolitical dynamics, and humanitarian and legal constraints. Technological advancements have revolutionised warfare introducing new complexities such as cyber capabilities, uncrewed systems, and artificial intelligence. The battlespace extends beyond the physical realm, with adversaries leveraging information warfare and cyber attacks. Dealing with these challenges requires advanced intelligence capabilities and robust cyber defence strategies.

Hybrid threats, which combine conventional and unconventional tactics, demand a comprehensive understanding of non-state actors and the ability to adapt swiftly. Military operations now span multiple domains, necessitating coordination and integration across land, sea, air, space, and cyberspace. Urban warfare scenarios and complex terrains add further intricacy to military operations, requiring specialised training and tactics.

Geopolitical dynamics significantly impact the modern battlespace, with territorial disputes and regional power struggles influencing military operations. Strategic decision-making must consider these complex considerations, fostering strategic partnerships and diplomatic efforts. Upholding international humanitarian law and ethical standards amidst the complexities of warfare is of paramount importance.

Effectively addressing and operating in the modern battlespace requires agility, adaptability, and innovation. Militaries must continually evolve strategies, invest in advanced capabilities, and foster joint and coalition partnerships. Embracing complexity means embracing change, harnessing technological advancements, and cultivating an organisational culture of adaptability.

The ability to navigate complexity will determine how effective a military is in achieving its mission and safeguarding national security. By understanding and embracing the challenges and opportunities presented by the modern battlespace, Western militaries can position themselves at the forefront of innovation and adaptability. Continuous learning and improvement are crucial to effectively responding to emerging threats and contributing to global peace and stability.

Complexity in the modern battlespace encompasses a wide range of challenges and uncertainties that the militaries of democratic nations must navigate. From technological advancements to hybrid threats, geopolitical dynamics, and humanitarian considerations, the modern battlespace requires

Figure 1.2 Shifting the Needle

agility, adaptability, and innovation. By embracing complexity and continuously evolving strategies and capabilities, militaries can effectively address emerging threats and contribute to national security in an ever-changing world (Figure 1.2).

1.3   Creativity, Design Thinking, and Innovation

Creativity, design thinking, and innovation are often used interchangeably, yet they have distinct meanings (synthesised from Taura & Nagai, 2017).

- **Creativity:** The cognitive process that forms new ideas and knowledge, involving the art of imagining what has never been seen before.
- **Design Thinking:** A user-centred process for shaping ideas and collaboratively solving problems.
- **Innovation:** The introduction of new ideas or processes that bring about change or improvement at tactical, operational, or strategic levels.

Design thinking serves as the intermediary between creativity and tangible realisation of innovation, acting as the conduit through which creative concepts are translated into practical solutions. It involves the deliberate shaping and organisation of elements to address specific problems or needs, guided by design thinking principles, which are characterised by empathy, iteration, and user-centricity (Bruton, 2011; Wrigley, 2017). Through design, creative ideas evolve from initial inspiration into structured frameworks and tangible forms, transforming abstract concepts into concrete innovative solutions (Figure 1.3).

The art of both creativity and innovation share many attributes. While there are many descriptors and, thus, ways to develop both, perhaps the biggest influence is freeing the mind of limitations. Of these, the most common include cognitive bias, heuristics, and habitual convergent thinking strategies. In contrast, to fully exploit the freedom to be creative or innovative artists, future military decision-makers need to be encouraged to develop divergent thinking skills.

This book delves into the emerging concept of military design thinking, an approach explored by many militaries worldwide and discussed in detail

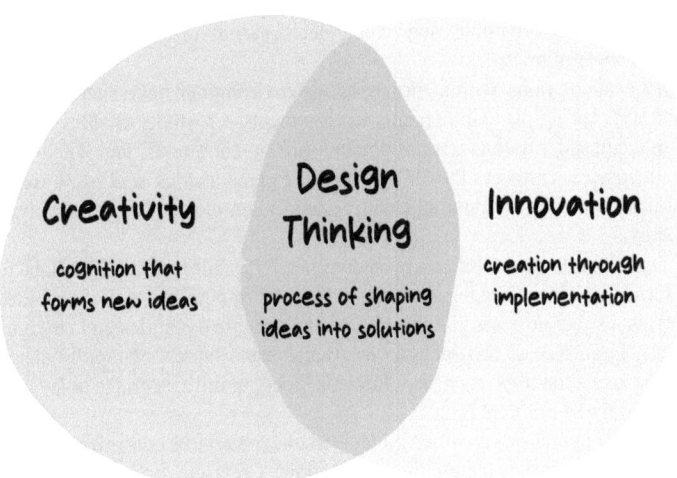

Figure 1.3 The Relationship Between Creativity, Design Thinking, and Innovation

in the following chapters. Military design thinking offers a pragmatic method for transforming traditional military problem-solving techniques, emphasising reflexivity and innovation to tackle contemporary challenges. This process thrives in uncertainty, with the ideation phase – often neglected in traditional planning – being crucial for generating a diverse array of ideas. By fostering and refining innovative solutions, military design thinking promotes a culture of continuous improvement.

1.4 Creativity and Cognitive Agility in the Military

Creativity, innovation, cognitive agility, and divergent thinking all play a crucial role in navigating military complexity. In the face of dynamic and unpredictable challenges, creativity enables militaries to generate innovative solutions, adapt to evolving situations, and gain a competitive advantage.

The need for unconventional approaches is omnipresent. Military operations often involve complex and ambiguous problems that cannot be addressed through traditional approaches. Creativity allows military personnel to think outside the box, explore new perspectives, and develop novel strategies to overcome obstacles and achieve mission objectives. For example, a cyber attack taking out all established secure communication systems will challenge local commanders to develop alternative forms. Operating within a dynamic and unpredictable environment, the military confronts strategic complexities, problem-solving dilemmas, and critical decision-making scenarios that shape the outcomes of missions.

Such adaptations within the changing environment necessitate military operations taking place in dynamic environments. Creativity enables militaries to adapt quickly to emerging threats, shifting conditions, and unexpected circumstances. It fosters the ability to adjust plans, tactics, and responses in real-time, ensuring operational effectiveness in complex and rapidly evolving situations.

Traditionally characterised by hierarchical structures and a penchant for uniformity, military organisations now face the imperative to embrace creativity as a strategic asset in light of evolving warfare dynamics and emerging threats. Examples of this include the trickle-down effects of rapidly changing political situations, such as politicians blocking military funds to leverage other unrelated political issues.

Military operations are often conducted under resource constraints, including time, personnel, and equipment limitations. Creativity helps maximise the efficient use of available resources, enabling innovative solutions and workarounds that optimise the effectiveness of military capabilities. Creativity amplifies operational effectiveness by nurturing adaptability and resilience. In the face of volatile and unpredictable circumstances, creative individuals possess the capacity to swiftly assess evolving situations and devise unorthodox tactics and strategies that outmanoeuvre adversaries. Examples of this include the transition of command control to mission command to empower local decision-makers scope to manage their own assets and request support for demands.

Military operations involve inherent uncertainties, where incomplete or ambiguous information can impact decision-making processes. Creativity allows military leaders and personnel to think critically, analyse available information, and develop creative strategies that consider multiple perspectives and potential outcomes, helping to mitigate risks and capitalise on opportunities. Today's military endeavours surpass conventional battlefields, propelled by asymmetric tactics, advanced technologies, and information warfare. In such a context, the cultivation of creativity assumes the utmost importance to enable military forces to navigate and respond effectively.

At its core, creativity empowers military personnel to envision novel strategies, exploit opportunities, and pre-emptively address nascent challenges. For example, since the 2022 invasion of Ukraine, the innovative use of drones forced both sides to develop counter-drone systems rapidly. Creative thinking is crucial for effective and adaptive leadership in complex military environments. Leaders must inspire and empower their teams to think creatively, in addition to encouraging diverse perspectives, open communication, and experimentation.

Creative leaders foster an environment that values intellectual agility, adaptability, and continuous learning. Creative problem-solving equips military leaders with the tools to surmount barriers, capitalise on advantages, and accomplish mission goals with greater efficacy – for example, the highly successful deception planning that preceded the D-Day Landings of 1944. This involved a complex mix of multiple activities, including decoy battle groups, washed-up dead couriers with fake plans, and a host of other subtle distractors. This example is not new, but it reinforces the point that great power competition demands complex innovation rather than the set-piece battlefield doctrine of smaller-scale wars.

The re-emerging threat of great power war needs more than just a rapid build-up of platforms. Even a cursory scan of the most innovative militaries would suggest they are the ones staring down the barrel of an invasion army. The imperative to innovate seems to enjoy a positive correlation with the level of existential threat. Conversely, invading forces who are engaged in wars of choice often appear much more complacent and less motivated to innovate. Many Western militaries have enjoyed a long period of relative peace and now seem ambivalent to the distant war drums beating on the horizon.

Though the threat of great power conflict does not feature in the living memory of the current generation, it demands a mindset characterised by openness, adaptability, and a willingness to challenge long-held assumptions. The infusion of creative thinking empowers military leaders and personnel to identify unconventional paths in overcoming obstacles; however, this comes with accepting the risk of failed attempts at innovation.

1.5 The Military's Innovation Dilemma

Since the beginning of recorded history, the value of strict discipline has been a hallmark of powerful militaries. Young and inexperienced personnel are often quickly thrust into harm's way and given access to equipment that can cause serious damage. It is imperative that militaries can rapidly prepare their newest members to not only remain safe around dangerous systems but also perform effectively while under immense stress. At the tactical level, in particular, internal conformity and predictability remain essential.

Pavlovian training regimes not only instil, but also maintain high levels of readiness for critical situation responses. This approach is common across most conventional militaries in the way they assimilate civilians into the organisation. Rigorous training with frequent repetition helps maintain this instinctive response behaviour – yet this typically comes at the expense of any divergent thinking encouragement. Eventually, though, with experience and recognised talent, junior personnel rise to leadership levels where they are progressively empowered to make decisions. Freedom to think for themselves, however, has now been diluted from their DNA.

Instinctive obedience and unquestioning conformity are both essential in critical situations. Indeed, the military has good reason not to encourage unchecked innovation. There are those who rightly caution against too much innovation, and thus, balancing this dynamic tension is key. Finkel (2019), for example, employs the term "over-innovation" when introducing his more measured concept of "conservatism by choice."

Junior ranks, despite their enthusiasm, often lack the bigger picture impacts of local innovation suggestions. The chain of command, therefore, reserves the right to limit their epiphanies to verbal proposals when invited. Over time, however, with experience and wisdom (typically indicated by increased rank), the system progressively empowers decision-makers to implement novel ideas. This unwritten spectrum spans oil and water at one end and wave-particle duality at the other, the lower level being where conformity and innovation are separated and the higher level, where conformity and innovation vacillate seamlessly.

Mid-level decision-makers combine their expertise prowess with deeper insights to consider new approaches to legacy systems. But some would question if the flame of genuine innovation has been snuffed out by this stage. Furthermore, career progression systems (Young, 2017) and fear of failure are also known to inhibit innovation. This problem is so well known that there are names for the phenomenon. "The frozen middle" (Jackson & Humble, 1994; Williamson, 2023) and "iron colonels" (Kalms & Sayer, 2020) are both used to describe those in their mid-career posts who are motivated more by protecting personal advancement than by organisational good. Professional military educational courses, such as pre-command, do not help this, with their steady stream of horror stories and legal officer briefs to strike fear into any ambitious innovator.

The diminishing relevance of legacy thinking is not lost on senior leadership. Strategic-level documents from the highest echelons regularly call for greater innovation, yet this seldom translates to noticeable organisational change. Both empirical and anecdotal evidence identify the organisational barriers that actively suppress innovative thinking. Despite the creation of dedicated innovation organisations, project teams (Parker, 2020), and the proliferation of "innovation" in job titles, the dominant organisational culture remains one of conformity and compliance – but not without justification.

The presence of some cultural resistance to innovation is further evidenced by self-stabilising terms to discourage divergent thinking. One example includes "situating the appreciation," which refers to the time-honoured tradition of predetermining an outcome before properly understanding the problem. This leads to incorrectly forcing templated solutions onto novel problems, regardless of how appropriate the proposed plan is. This peacetime shortcut approach is popular because it is seldom properly evaluated against a truly innovative adversary. Complacency becomes normalised when confidence is valued above competence in the leadership levels, despite the reverse being true at the lower ranks. A second example of anti-innovation is the phrase, "Don't fight the white." This age-old staff course axiom discourages students from thinking more broadly about a question and just conforming to what is expected.

Case Study 1: Vignette of Caution

To reinforce this chapter's purpose, it is crucial for senior military leaders to grasp the principles of complexity theory, as it offers valuable insights into navigating today's multifaceted operational environments. Embracing creativity in innovation is essential for adapting to these complexities, allowing military organisations to remain agile and effective in addressing the evolving challenges of modern warfare.

From the start of the 2022 Ukrainian War, civilian drone technicians found themselves mobilised to the front lines and rapidly tinkering with technology to outsmart the enemy. Being free of entrenched military mental models, they conceived ingenious ways to modify systems and exploit vulnerabilities in the adversary's Soviet-era capabilities. This rapid success wreaked havoc on the traditional-thinking invaders, who thought overwhelming force would deliver an immediate and decisive victory.

Encouraged by these early successes, the drone operators on the front line became even more adventurous. For some, it was not a big step to enhance these humble off-the-shelf drones to become autonomous weapons systems, but the target fixation of not getting killed was a tactical-level lens. These rapidly mobilised inventors were oblivious to the grand strategic-level challenges of securing international support aid. While technically not illegal under international law at the time, many countries which were heavily supporting Ukraine with military and financial aid might very easily have lost interest if the autonomous weapons redline was crossed.

The ingenious innovators in the trenches do not always appreciate the bigger picture. This is an example of knowing very clearly which rules (even unwritten ones) can be bent and which ones are written in blood. Meanwhile, at the highest level of government, where nurturing foreign aid is paramount, senior leadership needs to anticipate how their clever drone technicians on the front lines might be tempted. Thus, leading creative organisations demands not only an astute ability to anticipate future system states but also the foresight to pre-empt influences on undesirable states (Figure 1.4).

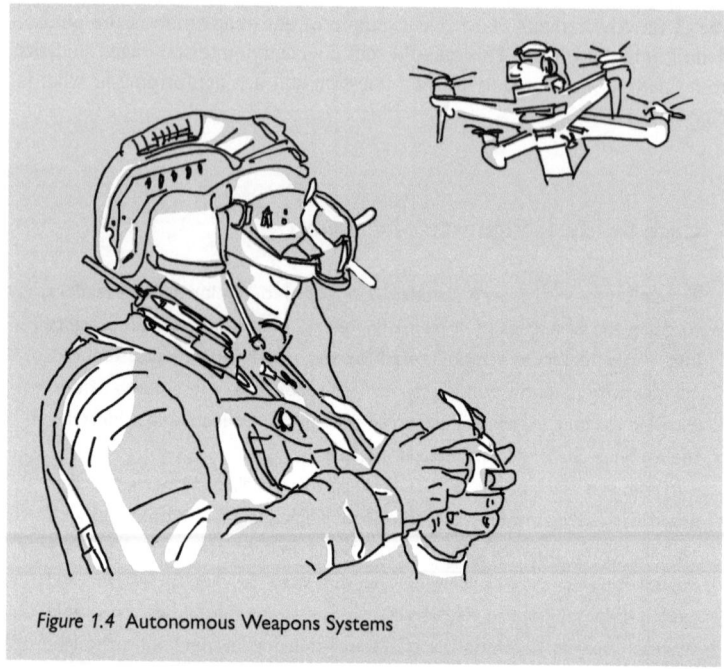

Figure 1.4 Autonomous Weapons Systems

1.6   Chapter Outlines

Against this backdrop, the forthcoming pages embark on an exploration of potential pathways forward to address this innovation dilemma within the military domain and delve into the challenging future awaiting all militaries, marked by their struggles to innovate.

Chapter 2: Systems Theory

This chapter offers an in-depth examination of the evolving intricacies within emerging battlespace environments. Given the dynamic nature of modern warfare, it becomes imperative for military personnel to adopt innovative approaches to strategic thinking. The chapter also discusses the Cynefin framework, which is used to assist military professionals in deciphering and navigating the multifaceted challenges posed by the operational context.

Chapter 3: Military Design Thinking

In this chapter, a critical evaluation of the limitations inherent in conventional military strategies amidst the rapidly changing face of warfare is presented. The significance of abductive reasoning and its reflexive role in fostering innovative and adaptable military solutions for contemporary challenges are discussed.

Chapter 4: Nurturing Creative Mindsets

This chapter explores the micro and macro perspectives of an innovative military culture. This is traced by considering inspired individuals, mavericks, and the silent majority who have the potential to foster this shift. The focus of the chapter then switches from a bottom-up, micro perspective to the macro-level levers that senior leadership can not only curate but also nurture.

Chapter 5: Military Organisational Constraints and Conditions

This chapter delves into the intricate barriers impeding innovation within military landscapes. By examining deep-seated biases and the prevalent apprehension surrounding potential errors, a comprehensive analysis of these impediments is presented. Furthermore, the chapter proposes methodical approaches and strategies designed to surmount these challenges, thereby fostering a more adaptive and innovative military environment.

Chapter 6: Towards New Horizons

In this concluding chapter, readers are encouraged to embark on a reflective journey, revisiting the insights and revelations unveiled in the preceding pages. By emphasising the criticality of perpetual innovation and fresh ideation amidst the dynamic complex military landscape, further exploration is advocated. However, it is imperative to note that this book does not serve as a replacement for localised recommendations aimed at crafting tailored action strategies. Instead, it offers an inspirational and thought-provoking narrative intended to equip leaders with a new lens through which to envision the future force and its requisites.

2 Systems Theory

Simplicity at the edges, complexity at the core.

2.1 Why National Security is Complex

National security and prosperity are complex, requiring our decision-makers to not only survive but also thrive in a dynamic environment. Paradoxically, the military is founded on a tradition of linear reductionist thinking where tactics, techniques, and procedures (TTPs) are both taught and documented in simple to complicated ways. Although this approach remains vital at the tactical level, there comes a time when simply doing things the way they have always been done is not enough. Particularly at the operational and strategic levels, decision-makers need to exploit alternative thinking methods if they are to successfully lead defence through the volatile, uncertain, complex, ambiguous, and novel (VUCAN) operating environment of the 21st century.

This chapter explores the shifting sands of national security issues and introduces options to better prepare decision-makers for this challenging responsibility. It begins by exploring emerging security trends that challenge traditional thinking models and underscores the need for fresh approaches. It then revisits the importance of both convergent and divergent thinking, integrating these concepts into some of the leading frameworks for addressing complexity.

Military leaders, from those devising the highest level of national security strategies down to those responsible for the execution of military responses, must possess a profound understanding of complexity theory. Warfare has always been inherently complex and chaotic. However, the advent of technology and globalisation has significantly shrunk the world, both in terms of time and interconnectedness. This transformation means that the challenges faced by military planners just two decades ago, though significant then, now seem relatively insignificant in comparison to today's rapidly evolving and unpredictable global security landscape.

DOI: 10.4324/9781003502180-2
This chapter has been made available under a CC-BY-NC-ND 4.0 license.

Perhaps the most overt shift in the military's contribution to national power has been the emergence of new domains. Not only do democratic nations still require traditional capabilities across land, sea, and air, but the increasing importance of space and cyber has seen these two reach the tipping point of becoming recognised domains in their own right. As increasingly seen in militaries around the world, there is a progressive shift from domain-centric, to joint, and now integrated. In the United States (US), despite its megalithic scale giving institutional inertia, the Goldwater Nichols Act (1986) directed increased joinery. Both the new domains and the pragmatic realities of *integrated* over *joint* are yet to mature, and their impact is yet to be fully tested. Cultures are hard to change.

Strategic interests and national security objectives guide a country's approach to the contemporary battlespace. These objectives may include preserving national sovereignty, protecting vital maritime trade routes, fostering regional stability, and contributing to international peacekeeping and humanitarian efforts. To effectively address these challenges, the West must maintain a credible and adaptable defence posture that can respond to a wide range of security scenarios.

At the operational level, military planners also need to reinvent the way they think, not only with the formal establishment of new domains and the shift to integration at the organisational level but also with how these translate into military responses. In navigating the modern battlespace, the West must consider a range of influences. Regional dynamics, such as shifting power balances and territorial disputes, play a significant role. Global power shifts, including the rise of new actors and evolving alliances, further shape the strategic landscape. Technological advancements introduce both opportunities and vulnerabilities, requiring militaries to adapt and keep pace with the changing nature of warfare. Additionally, non-traditional security threats, such as terrorism, organised crime, and climate change, continue to add complexity to the battlespace.

Within the modern battlespace, several key considerations come to the forefront. Developing advanced military capabilities becomes imperative to deter potential adversaries and safeguard national security. Robust intelligence and surveillance systems enable situational awareness and early detection of threats. Effective command-and-control structures facilitate the coordination and execution of military operations across multiple domains. Yet the modern battlespace extends beyond conventional military operations. Hybrid threats, which combine both conventional and unconventional methods, pose complex challenges. Asymmetric warfare tactics, cyber attacks, and information warfare require increased attention to cybersecurity, resilience, and the protection of critical infrastructure.

The contemporary battlespace is characterised by its dynamic and ever-evolving nature. For Western militaries, the imperative of continuous adaptation and investment in state-of-the-art technologies are paramount, coupled with a profound comprehension of the shifting spectrum of threats and opportunities. This chapter explores how militaries – traditionally encumbered by conventional methodologies – can transcend such confines. It advocates for innovative defence strategies that not only ensure national security and the safeguarding of strategic interests but also facilitate a significant role in enhancing regional and global stability. The focus is on empowering senior military and industry leaders to overcome the notorious "valley of death" by fostering cognitive agility and bold decision-making, turning these concepts from mere buzzwords into practical, actionable strategies.

2.2   The Necessity of Thinking Differently

In democratic nations, the traditional role of the military is to uphold stability. For most, this typically involves ensuring peace and security, thereby facilitating the nation's pursuit of prosperity. This role includes deterring emerging threats and, when necessary, responding effectively to disrupt or neutralise them. The ultimate aim is to restore the country to its prior state of stability, often referred to as the *status quo ante*.

In reality, history has shown most Western militaries to be more expeditionary than defensive by either responding to the call of their allies or helping promote global stability through the rules-based international order (RBIO). Throughout the past century, however, such contributions have been largely military-led. In fact, "many Western armed forces promoted the operational level as a space where military expertise could be left alone to solve military concerns" (Carr, 2024, p. 125). The problem here is that from enlistment up, decision-making techniques have been built on a foundation of treating situations as merely complicated. Thus, set-piece tactics, techniques, and procedures (TTPs) are blended with rote-learnt doctrinal models and checklists – all of which have been retroductively derived from previous operations. While failures are removed, even previous successes need to be contextualised through the zeitgeist of the unique environment. Studying military history needs to be so much more than just reviewing wars; it must help derive enduring principles of warfare, but more importantly, it must also enhance cognitive agility to extrapolate into the new and *sui generis* future. Military commanders do not have time to develop innovative thinking once the bullets start flying. By then, it is too late, and failure is not an option – or, to paraphrase German Chancellor von Bethmann-Hollweg, "diplomacy ends when the iron dice roll."

Entrenched military mindsets have their place. The need to recruit young, inexperienced personnel and have them combat-ready in minimal time is enshrined in most initial military training regimes. Strict discipline and simple procedures remain vital for both keeping people safe and achieving victory on the battlefield. The extreme *cognitive load* (Sweller, 2010) pressures of high-stress environments mean the military must absolutely drill basic procedures into its most junior personnel. Such conditioned responses must be habitual and instinctive.

The military is renowned for strict order and discipline, so much so that they are called in to help restore order during natural disasters, such as earthquakes, floods, and pandemics. The combined effect of repetitive drilling of set tactical procedures and the linear operational planning process means that the military excels in convergent thinking. For most of their careers, military personnel are both taught and rewarded for quick decision-making and restoring order. This, however, is achieved by modifying pre-existing templates to current situations, a technique that works well in lower levels where multi-order consequences are managed or tolerated but becomes problematic higher up.

Self-stabilising is a feature of complex adaptive systems where equilibrium is restored when destabilising inputs act on the system. Not only does the military have well-rehearsed responses to known situations, but it also often discourages free thinkers who deviate from the norms. The combination of these forces means the military has a strong organisational culture of convergent thinking. At both the individual and collective level, there is an instinctive sense of "running to the sound of the guns." This rush for a solution can, however, often come at the expense of taking time to understand the situation fully. The problem is so prevalent that it is even given a name: "situating the appreciation."

The prevailing culture of convergent thinking is further enhanced by military recruiting strategies. The overt emphasis on science, technology, engineering, and mathematics (STEM) backgrounds comes from two main drivers. While attracting new recruits is perceived to be more successful when STEM qualifications are offered, there is also a misconception that defence needs more STEM graduates in their decision-making ranks. And while it is true that STEM is important for many fields, it does prioritise convergent over divergent thinking. In fact, the recruiting instruments for selecting officer candidates are skewed in favour of IQ tests where pattern recognition is required. This convergent thinking attribute is clearly valuable for STEM roles and, indeed, the tactical levels of compliance, but not in the mid-career world of complex adaptive systems.

The recruiting, developing, and rewarding of convergent thinking comes at a cost. Unlike industry, militaries cannot easily "go to market" to employ additional qualified staff who are deeply assimilated in the ideology of the profession of arms. Nonetheless, select roles can indeed be recruited direct from market for specialist deployable appointments or simply employed as civilian staff. The unique nature of deploying in harm's way with unconditional service means practitioners must be homegrown – but with selected corporate sector models and ideas adapted for use. In fact, military decision-makers need to be well educated in the best ideas from industry to know what ideas can be employed in military contexts. Ultimately, though, the military does not make toasters.

The junior echelons are the breeding grounds of future key decision-makers. Through self-selection and many years of reinforcement in convergent thinking, those who succeed in the incentivised convergent thinking system are then thrust into the uncomfortable world of needing to exploit divergent thinking. Reigniting dormant thinking skills is not a quick process and is almost impossible if those individuals who have excelled in being naturally one type over the other.

The culture of convergent thinking is exacerbated by the fact that the occasional divergent thinkers who slip past the recruiting psychologists often become disillusioned by the organisation's obsession with compliance. In a phenomenon known as *last person standing*, the organisation is said to *eat its young* by weeding out the innovators in their junior years. The others who bubble to the top are those who crave predictability and the certainty of convergent thinking. They become the approvers of recognition, reward, and promotion systems, which in turn makes it a self-perpetuating system. This reinforces the culture of convergent thinking with artefacts that promote self-stabilising influences (Young, 2017).

Structuring the organisation to preserve the historical culture of training dogma is only part of the problem. Beyond preserving the promotion systems to ensure their successors think just like them, there is a mutual appreciation–clique effect where like-minded leaders subconsciously promote tacit endorsement of their peers who think and act like them. Psychologists call this intra-group attraction social identity theory (Ellemers & Haslam, 2012), while less complimentary critics describe such members as *iron colonels.*

Many Western militaries also suffer further anti-innovation forces in the guise of fear of failure. Linked to the skewed promotion systems and the social conformity pressures of cronyism is the pervasive culture of survival. The pinnacle of most officers' careers is their time in command. For those who even achieve this milestone, it is generally a single-shot opportunity and a defining moment in establishing their reputation. While many heirs-apparent aspire to achieve great things during their moment of fame, the fear of failure seems to dominate. This problem is such a widely recognised phenomenon that the chattering class refers to them as the "frozen middle" (Williamson, 2023). But like so many other blockers of innovation, these people are not deliberately seeking to be anti-innovation. Their pre-command courses are stacked with horror stories and draconian threats of eternal consequences that all but guarantee commanders will fall into the risk-averse "not-on-my-watch" syndrome. Almost ironically, and despite knowing of this problem, hardly any budding commander actually wants to perpetuate the entropic death of the frozen middle.

All militaries have not only divergent thinkers but also celebrated histories of ingenious innovations in the workplace. Countless job titles include the word innovation, and there are a number of dedicated entities whose sole job is to find and support innovators. Most militaries also have a sizeable contribution of civilian staff who are not subjected to the operant conditioning of negative reinforcement endemic to formal military courses. So, while still assimilated into the pervasive workplace culture of entrenched conformity, these civilian staff are slightly less vulnerable to convergent thinking. Furthermore, most militaries have dedicated organisations of civilian scientists (e.g., Defense Advanced Research Projects Agency [DARPA], Advanced Research Projects Agency–Energy [ARPA-E], Defence Science and Technology Group [DSTG], etc.) who exist to provide both contestability and alternative thinking to challenging problems (Vallerand & Masys, 2022). Most, however, are predominantly STEM oriented and, like other defence civilian staff, not the ones who must negotiate the VUCAN world of national military power. The real problem, therefore, is to empower combat-ready uniformed decision-makers who exude innovative mindsets when dealing with wicked systems – not just miniaturising exquisite technologies, such as exploding pens. Synchronic product design and diachronic systemic design are not the same thing.

2.2.1   *Types of Thinking*

Convergent and divergent thinking are both important when dealing with complexity. Although their relative merits will always be debated, there is a need to assist defence personnel in having a balance of both to ensure that decisive solutions can be found when appropriate and innovative ideas can be exploited for complex situations.

Divergent thinking is a thought process or method used to generate creative ideas by exploring many possible solutions. It is nonlinear and not structured by predetermined rules or patterns (Figure 2.1). This kind of thinking involves free-flowing thoughts and is often linked with creativity because it encourages the exploration of many new, diverse, and even seemingly disconnected ideas.

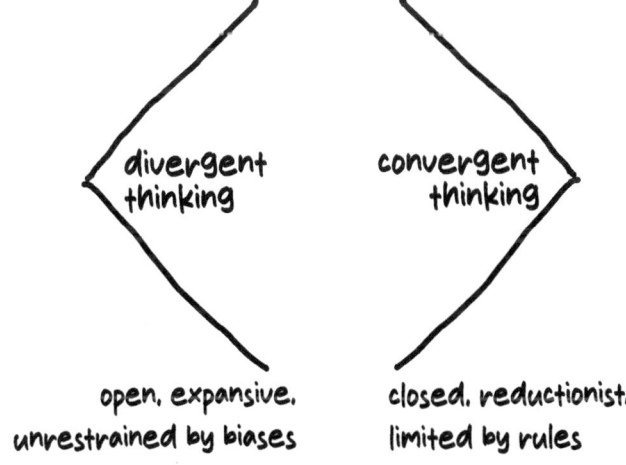

Figure 2.1 Divergent and Convergent Thinking

In contrast, convergent thinking is systematic and logical. It involves following a set of coherent steps to arrive at one correct solution. It's the type of thinking one uses when solving a maths problem, for example. Convergent thinking is characterised by the ability to give the "correct" answer to standard questions that do not require significant creativity.

In many military organisations, convergent thinking has historically been emphasised because it aligns with the need for uniformity, hierarchy, and discipline. Convergent thinking leads to quick decision-making, which is crucial in many military situations where time is of the essence, and the cost of failure can be extraordinarily high. This thinking style supports the enforcement of standard operating procedures and protocols that ensure consistency and reliability in high-stakes environments.

However, the preference for convergent thinking can suppress divergent thinking, which is equally important, especially in modern warfare and peacekeeping operations where unpredictability is a constant. The ability to think divergently allows for flexible adaptation to new threats, creative problem-solving, and innovation in tactics and strategy. When divergent thinking is discouraged, the military risks becoming stagnant, less adaptive, and potentially unable to think "outside the box" to counter non-traditional threats or devise novel solutions to complex problems.

Divergent thinking is crucial for the evolution of military tactics and the integration of new technologies and methodologies into defence protocols. In the rapidly changing landscape of global security, fostering a balance between convergent and divergent thinking can be a significant force multiplier. The challenge for the military and similar organisations is to cultivate an environment where the regimented, disciplined approach that is necessary for military operations can coexist with and is enhanced by the creativity and innovation that divergent thinking brings.

Addressing the challenges of complex adaptive systems demands both convergent and divergent thinking approaches. While convergent thinking employs reductionism to find a single right answer (positivism), divergent thinking requires a more expansive exploration of opportunities (post-positivism). Convergent thinking is perfect for complicated problems and is a respected approach when the situation is accurately recognised. The challenge is developing the cognitive skills to recognise when a situation is more complex than complicated. Once recognised, a decision-maker needs to take care not to misemploy the wrong approach. Shoehorning complicated "solutions" on complex systems can lead to disastrous consequences.

2.2.2   Epistemic Diversity

If you want to go fast, go alone.
If you want to go strong, go together.

– A paremiological mystery

Epistemic diversity speaks to the importance of multiple *ways to think* about a problem. In the case of systems design, this means having interdisciplinary representation at the table. Both the emergent nature of warfare and the intricate complexities of contemporary security environments demand that military personnel engage in divergent thinking. This diversity is justified by several key factors that highlight the need for fresh perspectives and innovative approaches:

Complexity and uncertainty: While military commanders have always faced their own complex challenges, today's landscape is transformed by the reach and velocity of technological advancements (e.g., real-time social media impacts on the battlespace). These developments have drastically compressed decision-making cycles. Adversaries now often resort to unconventional tactics and employ hybrid warfare strategies, leading to highly fluid and uncertain battlefields. In this context, divergent thinking becomes indispensable. It empowers military personnel to view problems from varied perspectives, conceive innovative solutions, and adeptly manoeuvre through these intricate and unpredictable scenarios. The ability to think divergently is no longer just an advantage but a necessity for effectively addressing the rapidly evolving nature of modern warfare.

Technological advancements: Throughout history, the advent of new technology has consistently reshaped military strategies and tactics. However, what sets the current era apart is the exponential rate at which technological progress is occurring – a pace unlike anything witnessed before. This rapid evolution presents a unique blend of opportunities and challenges for the military. Contemporary developments, from cyberwarfare to artificial intelligence, not only require a paradigm shift in thinking but also demand the agility to embrace and integrate new capabilities swiftly. Divergent thinking is crucial in this context as it enables military personnel to explore and understand the potential applications of emerging technologies thoroughly. By leveraging this mindset, they can effectively exploit technological advantages and adapt to the ever-changing landscape of warfare, where the rapidity and scale of innovation have become defining characteristics.

Multi-domain operations: Contemporary military operations intertwine all five domains (land, sea, air, cyber, and space) in dynamic and real-time ways never seen before. Efficient coordination and integration across these domains require a holistic and innovative approach. Divergent thinking encourages a broader perspective, enabling military personnel to recognise interconnections, exploit synergies, and develop integrated solutions across diverse operational domains.

Adversarial adaptation: Adversaries constantly adapt their strategies and tactics to exploit vulnerabilities and counter conventional military approaches. The modern incarnation of John Boyd's famous observe, orient, decide, and act (OODA) loop of lessons learned (lessons noted) is often real-time. To maintain a competitive edge, military personnel must anticipate and understand these adaptations, devising effective counter-measures faster than the adversary. Divergent thinking facilitates the creation of unconventional strategies and techniques that can outmanoeuvre adversaries and provide asymmetrical advantages.

Innovation and progress: The military must stay at the forefront of emerging threats and technological advancements. Cultivating diverse thinking promotes innovation, fosters exploration of new concepts and technologies, and facilitates the development of cutting-edge capabilities. The historic concept to operational lead times needs to shrink faster than the adversary can react. Divergent thinking drives progress, ensuring that the military remains effective and resilient in an ever-evolving security landscape.

2.2.3   Is the System Broken?

Complex adaptive systems are never broken; they just are. Unlike complicated ones, where a single right answer can be deduced, complex adaptive systems ebb and flow with shifting influences. Different observers will have their own opinions of what parts of the system are better than others, but ultimately, complex adaptive systems are merely perceived to be in more or less favourable states at any given time by any given observer. The degree of military innovation will be assessed differently by everyone – as with the perception of whether it is too much or not enough.

Innovation is only a good thing at the right time and place. The military needs stability and predictability as well, so the two are arguably in a constant state of dynamic tension. There are those who actively discourage divergent thinking in their particular work area – and for good reason. In contrast, senior leadership frequently calls for improved innovation across the organisation. While innovation exists, its effectiveness hinges on three key factors: the degree to which it meets critical needs, whether staff understands their empowerment to innovate, and their awareness of the organisation's tolerance for failure in specific circumstances.

Knowing when the level of innovation is good enough is problematic. While the military's prowess is obviously tested on the battlefield, where tactical winners and losers are more obvious, this is not necessarily the best evaluation. Many would argue that the true job of a military is actually to prevent wars, not win them. *Shape* and *deter* are, in fact, a 24/7 task; *responding* is only required when the other two have failed. Strategic success, therefore, is about maintaining the peace, not winning the war.

The most celebrated examples of military innovation occur in combat. Compared with the overly bureaucratic constraints of peacetime, once the *iron dice roll*, the military enjoys significantly more freedom to take risks and explore innovative courses of action. The art here is to forge a military that can successfully pivot to divergent thinking despite being incubated for so long during the peacetime obsession with convergent thinking.

The balance of military and corporate mindsets needs careful consideration. Although engaging civilian staff for non-deployable roles is fiscally prudent, and funding military personnel to study in civilian business schools is also pragmatic, caution is necessary. Many traditional business concepts are still heavily influenced by profit-linked KPIs, and competitive practices are motivated by hyper-efficiency and minimal wastage. While it is true there has been a shift, especially since the COVID-19 pandemic, notions of just-in-time logistics and task-specific professional development are entirely appropriate for businesses where they can simply recruit or buy in shortfalls when required. For the military, however, stockpiling war reserves and just-in-case training for unlikely situations is not just fundamental to defending the nation during its darkest hours; it is actually a deterrence against needing to respond. The military's need for *just-because training* (character development, antifragility, courage under fire, etc.) is vital when preparing personnel for the extreme psychological hardships of combat, but it would be difficult to find "just because" in any corporate textbook. Thus, while military decision-makers must keep a watching eye on civilian best practices, an unexamined copy-and-paste approach warrants caution. Yet this does not negate the value of reading widely.

Transient advantage is more than just having a competitive edge over an adversary. While the cat-and-mouse challenge of outmanoeuvring a potential enemy's technological advancements might seem obvious enough, the real test is unknown. The notion of transient advantage refers to the complex adaptive system of dynamic relativity. Although an edge might be achieved in one area, others might degrade. Despite this being highly insightful, there is a danger isolated threats are treated as equal. This reduces the threats to numbers on a page, operating in singular time and space without taking into consideration the holistic system's interplay. As complex adaptive systems, these entangled technologies can potentially develop levels of emergence to create unforeseen nonlinear tipping points that create a strategic shock.

The true effectiveness of a military system can only be fully assessed when it is tested in an actual engagement against a peer or near-peer (great power) adversary. This reality, however, makes the art of determining when a system is "good enough" quite challenging. While innovation and the development of new gadgets and ideas are celebrated achievements, merely counting these advancements doesn't necessarily equate to success. Without a comparative benchmark, such as data from a control group, relying solely on a one-sided tally of achievements is insufficient. It's akin to the sound of one hand clapping – an incomplete measure. In terms of formal logic, basing success exclusively on the number of innovations is a non sequitur argument; it's a flawed conclusion that doesn't follow from the premise and highlights the very cognitive biases and heuristics that are typically discouraged. A comprehensive evaluation requires a more nuanced approach, taking into account not just the innovations themselves but their effectiveness in real-world scenarios involving complex adaptive systems.

Case Study 2: Survivorship Bias

During World War II, a classified program was set up to enhance the survivability of aircraft under enemy fire. The initial observations were based on aircraft that had returned from missions, often riddled with bullet holes (Figure 2.2). The intention was straightforward: to determine where heavy armour plating was most needed to protect the aircraft. A team of mathematicians meticulously analysed extensive datasets detailing the bullet holes' locations on these aircraft. Their initial findings indicated a concentration of bullet holes in the fuselage and wings rather than around the engines, leading to a preliminary recommendation to reinforce these areas.

However, this approach was based on a false and unexamined assumption – that the aircraft that returned were representative of all aircraft.

Figure 2.2 Average Distribution of Bullet Holes on Returning Aircraft. Adapted from: Trevor Bragdon (2017)

The breakthrough came when statistician Abraham Wald pointed out a significant bias in the dataset: it only included aircraft that had survived their missions. The missing, yet vital, piece of the puzzle was the aircraft that didn't make it back. Wald reasoned that the areas not showing significant damage on the returning aircraft were actually the most vulnerable. If the engines and cockpits had been hit, the aircraft were more likely to be lost and thus not represented in the data. His insight led to a counterintuitive, yet essential, recommendation: to armour the seemingly less-hit areas like engines and cockpits.

Survivorship bias, where conclusions are drawn from an incomplete set of data that only includes "survivors." It underscores the importance of challenging initial assumptions and perceived knowledge, especially in complex problem-solving scenarios. By questioning the validity of their data source, the team was able to correctly identify the problem and provide a solution that significantly improved aircraft survivability. This story from World War II remains a powerful reminder of the necessity to rigorously examine the assumptions and the data used to ensure we are addressing the right challenges with effective interventions (Ellenberg, 2014).

This case study not only highlights survivorship bias but also points to the dangers of relying on incorrect or underrepresented datasets. This scenario is akin to the anecdote of a drunk man searching for his keys under a streetlamp. When a passer-by inquires about his actions, the drunk reveals that he actually dropped his keys up the road but is searching under the lamp because that's where the light is. This story illustrates the tendency to look for answers where it's easiest rather than where they're most likely to be found.

The concept of spurious correlations further complicates data analysis. For instance, though there might be a statistical link between increased shark attacks and higher ice cream sales on certain days, this doesn't imply a causal relationship. This is a classic example of the logical fallacy *post hoc ergo propter hoc* (correlation does not imply causation).

Applying this logic to other scenarios, such as evaluating the number of gunshot victims who sustain leg wounds compared to chest wounds, can be misleading. In the same vein, the number of innovations celebrated within the military doesn't necessarily indicate a strong culture of innovation. This raises the question: When does a certain amount of something become significant enough? A more pertinent question regarding survivorship in a military context might be whether defence is sustaining a transient advantage over potential adversaries.

2.3   Rebalancing Thinking Styles

To thrive in complex adaptive systems, defence needs to rebalance its over-emphasis on convergent thinking. This is not to say that convergent thinking should be neglected, but greater value can be afforded to also developing divergent thinking. Commensurate with encouraging both types is the need to help defence personnel recognise when the best time is to use one over the other. Before exploring the specifics of how to develop each of these, it is worth considering the leading theories on dealing with complexity.

2.3.1   Problem Types

Problems come in many shapes and sizes. Most people readily identify a *simple* problem as being a problem with just two parts, and their relationship is both obvious and predicable. Complicated and complex, however, are often incorrectly used interchangeably. Complicated is merely an extension of simple but with more parts. As a system increases in the number of components, it becomes more complicated, but the relationship between the components remains predicable and knowable. By using flow charts, wire diagrams, algorithms, or logic gates, an extremely complicated system can still be reduced down to discrete subsystems, making single outcomes possible. A complex system, however, also has multiple parts, but the interdependency of these elements means no single path can be drawn to solve an exact outcome. Complex systems are highly interdependent, meaning the possible outcomes can be circular and impossible to determine. Social media is a good example of a complex system.

Complex adaptive systems refer to the dynamic interplay of self-stabilising and emergent properties. Many complex systems have features that allow them to restore stability by freezing or reversing influences that attempt to change their state. Traffic lights, for example, help resist attempts to destabilise the system by maintaining orderly flow in busy cities. Meanwhile, other attributes of complex adaptive systems tolerate or even enhance inputs that influence the equilibrium. The impact of drought, for example, could lead to an animal herd moving to new feeding grounds. The catastrophic impact of climate change, however, might be too fast for evolutionary change for some species and thus lead to their extinction.

The notion of complex systems can be traced back to the earliest philosophers (Aristotle's *Holism* and *Organon*) and Leonhard Euler's network theory (1741, p. 92). The more modern interest, however, falls under the title of general systems theory (von Bertalanffy, 1968). This spawned a wave of other theories and interest around the fragility of industrial-era reductionism (Figure 2.3). In fact, the US military's creation of what became the internet (ARPANET) might not have been specifically designed to mitigate *single points of failure* in what was previously treated as just a complicated structure,

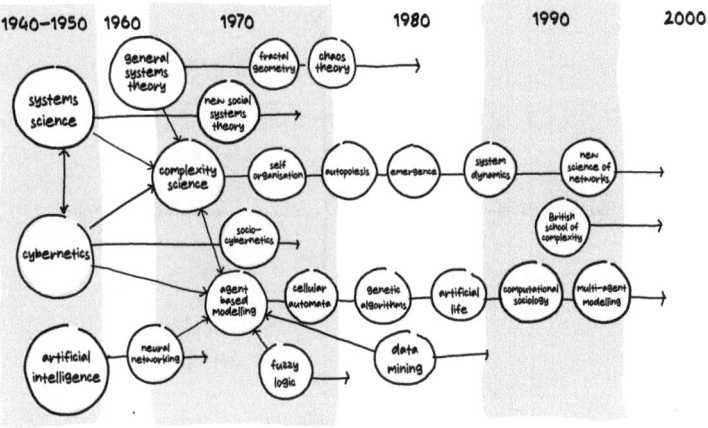

Figure 2.3 Military Complexity Has Been Recognised for Decades

but it certainly facilitated system resilience. Today, the understanding and exploitation of complexity theory continue to distinguish the progressives from the Luddites.

Beyond the military, civilian academics and practitioners alike have also contributed to the discussion on how best to excel in the realities of global security. Understanding and shaping complex systems is about exploiting complexity to make the adversaries' job harder, not just about making Blue Force operations easier. VUCAN works both ways in that it is not the problem; rather, the problem is not being able to deal with VUCAN.

2.3.2 Dealing with Complexity is Relatively Simple

Numerous theories have been developed to address the challenges of complex systems. One of the earliest and leading theorists is Russel Ackoff (1981, 1994, 2015), but others include Zwicky (1967), Churchman (1971), Rittel and Webber (1973), White (1975), Tukey (1977), and Pidd (1997). More recently, the Thomas-Kilmann conflict management model (McPheat, 2022) and Harvard Business School's cultural profile model (Groysberg et al., 2018) have become instructive. The latter shows how organisations can map their current and desired in-use culture on a plot comparing complicated to complex against self-stabilising to emergent dimensions (Figure 2.4).

Each of these theories has contributed to our growing understanding of not just different problem types but ways of dealing with complex, complex adaptive, wicked, chaotic, disordered, messy, and super crises. Although this book cannot provide detailed summaries of each, there is value in discussing the

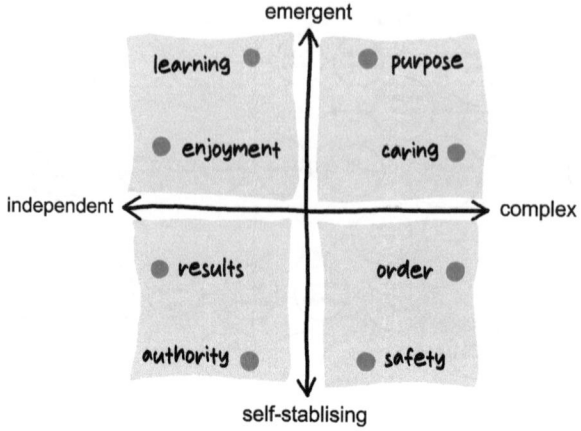

Figure 2.4 A Variation of the Cultural Profile Model. Adapted from: Groysberg et al. (2018)

leading concepts. The first of these is Russel Ackoff's (1981) seminal treatise (Figure 2.5).

Ackoff offers four broad approaches to dealing with problems: *absolve, solve, resolve*, and *dissolve*. The first is *absolving*, although some would argue that this does not constitute an actual approach. Absolving means to ignore a situation in the expectation that it will self-heal or become irrelevant. A typical example of this is when politicians avoid, distract, or deflect media attention regarding a negative event in the confidence that the news cycle will soon become fascinated by something else. While this strategy can lead to unwarranted consequences, risk mitigation is often factored into the decision to ignore the problem.

Solving is the most obvious and well-known approach to dealing with problems. Solving is the application of established principles and methods to determine a single correct answer. Solving works well for simple to complicated problems where all the variables are known, and the outcome is knowable. The more complicated the problem, the more complicated the process needed to solve the problem. Mathematical problems are a good example.

Solving is a transactional-level activity. The steps required can be tested and refined over time to either improve or update to suit changing circumstances. Furthermore, because complicated problems can be reduced to algorithmic formulas, they can be solved with computer programs. Typical examples of these include course plotting for naval vessels or aircraft and firing solutions for gunnery. Each of these was once the art of highly trained experts, but today are confined to nostalgic stories over brandy. While some complicated problems still require solving by specialists, they are increasingly being overtaken by technology. Any military decision-maker who can be replaced by a computer—deserves to be.

Resolving is the most commonly employed technique for true decision-makers. This approach is also known as "satisficing" and means accepting compromises to achieve "good enough" outcomes. This approach is particularly necessary when resources are insufficient to solve the problem. Resources can include anything from time, energy, personnel, materiel, data, or even the decision-maker's cognitive capacity. It relies on mental models and pattern recognition to make a best fit based on previous situations.

Resolving is the military's go-to approach when teaching decision-makers to cope with VUCAN situations. This process begins by instilling multiple templated processes for various scenarios. When a decision-maker is confronted with a new or unexpected situation, they choose the nearest cognitive schema to suit the situation and continue to apply it until it no longer helps. At this point, they reassess the situation and attempt to either change to another pre-programmed schema or modify it by combining several schemata. This process of primarily convergent thinking, mixed with occasional divergent adaptations, is a great developmental approach for helping decision-makers move from transactional to transformational outcomes. A good example is when a first aider arrives at a multi-vehicle crash and needs to apply first aid. The actual scene is quite different from the comfortable classroom setting where they learnt first aid, so they do the best they can to apply what has been taught but make compromises when necessary. Psychologists call this process *cognitive equilibrium* (Paiget, 1918).

Resolving problems has the potential to make situations worse. While it gives the appearance of fixing the immediate problem, it typically leads to undesirable consequences. In the heat of the moment, military decision-makers are primarily focused on the problem in front of them (e.g., just stop the bleeding). Intrinsic and germane cognitive load generally reduces the decision-maker's ability to foresee the negative consequences. When the situation is actually complex, rather than just a more complicated one than they have learnt to deal with, the resolution will have consequences. Sometimes, these consequences are sufficiently negligible, or their time delay will allow follow-on interventions to be developed, or the impact will be an acceptable risk. Resolving can be a viable technique, but it can also be problematic.

Resolving can be employed in both complicated and complex problems where high uncertainty exists; in other words, in multi-criteria decision-making with partial or missing datasets and possibly conflicting options (i.e., dilemma, trilemma, black swan, and super-crisis). While complex problems are almost impossible to mimic, computer modelling and simulation can help with complicated problem optimisation. For example, determining a submarine's maintenance activity cycle can be hampered by an unexpected supply issue. When extrapolated out to fleet-level scheduling, the number of variables exceeds the cognitive capacity of normal solving techniques (Marchau et al., 2019; Turan et al., 2021).

Computer modelling can be a helpful aid in exploring options for complex adaptive systems. Standard mathematical approaches deal with *deterministic predicable systems* (complicated) and *deterministic chaotic systems* (complicated but unable to be accurately modelled). However, more sophisticated use of *automatic differentiation* modelling using forward and backward Euler methods with Jacobian matrices can be used for the random probability of *stochastic systems* (Petty & O'Byrne, 2024), such as the submarine example above. Such modelling usually takes time to develop but can be a decision support tool to complement military commanders' innovative thinking.

While resolving can be used for both complicated problems and complex systems, the risk of catastrophic consequences increases significantly for the latter. Complicated problems can be reduced and isolated subsystems treated separately. Complex ones, however, require a holistic consideration of the entire system. A hospital patient with a sore knee could be offered painkillers and sent away; however, if the actual cause of the injury is not determined and rectified, the injury might get worse. An operational plan that successfully defeats an adversary might be considered a success, yet the way the apparent victory is achieved may ferment longer-term ideological issues that subsequently resurface on an even worse scale. The military axiom "winning the peace, not the war" speaks to the recognised problem of resolving a conflict.

The military has been operating in complex adaptive environments for millennia, and resolving has been their default approach. While the consequences of such action have either been brushed aside as collateral damage or ruefully lamented, there are other ways they mitigate the risk, one of which is to restrict the number of elements in a system. For example, inexperienced junior military leaders will only be given small teams of personnel with limited objectives. By constraining the number of elements, the subsystem might still be complex, but the potential for unexpected consequences is reduced. For example, leading a small squad to clear a room in a building can be a highly rehearsed tactic that allows for fairly predictable variations simply because of the reduced number of elements. As decision-makers grow in experience and build greater schemata, they are progressively entrusted to confront more complex challenges.

Above the operational level of warfare, the number of moving parts in a system becomes more than the working memory can manage. With the addition of multi-agency operations or coalition partners – who bring their own schemata – the degree of complexity becomes untenable. Yet the consummate military professional has been raised on a diet of transactional thinking strategies blended into a resolution mindset for variations. Furthermore, they have risen above their peers for thinking fast and demonstrating decisive leadership. Those who are the "best of breed" for treating everything as merely complicated are often the worst decision-makers for recognising when a system is actually complex. Continuing to employ resolution techniques on isolated parts of a system means the holistic situation remains neglected. Those fast-thinking skills that "got them here won't get them there." Thinking fast is about exploiting schemata, but this brings the darker side of cognitive bias and heuristics (Kahneman, 2011). The more drilled the decision-maker is in exploiting entrenched schemata, the more vulnerable they are to mismatch errors.

Complex adaptive systems require a fundamentally different mindset. As junior decision-makers transition from bounded complicated problems up to larger-scale unbounded ones, they shift from solving (prescriptive methods) to resolving (principle-based approaches). Further experience and rank progression, however, involves not only greater empowerment but also increasingly complex and adaptive challenges. This, in turn, requires a paradigmatic mindset shift to Ackoff's notion of *dissolving*. As will be explored in later chapters, the key aspects of this approach are captured within the broad field of design thinking. This involves the deliberate mitigation of cognitive biases through a diversity of interdisciplinary teams and various workshop activities to encourage divergent thinking. Other key attributes include heightened empathy through personas and semi-structured processes to crosscheck for cognitive blind spots.

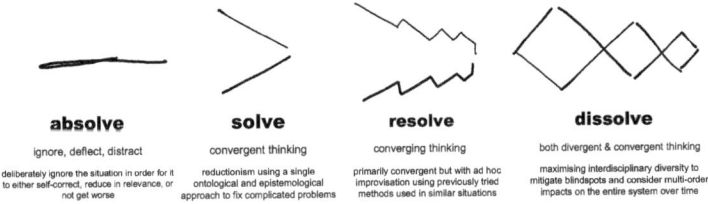

Figure 2.5 Ackoff's Theory: Absolve, Solve, Resolve, and Dissolve

While multiple different design thinking models should and do exist, the general concept is one of vacillating divergent and convergent thinking. This will be discussed more in Chapter 3; however, the general concept captures the essence of exploiting both thinking styles to progressively refine options for targeting the most challenging of situations.

2.4   Cynefin Framework

The Cynefin framework is another well-known and popular model for highlighting the need to treat problem types differently. Snowden and Boone (2007) model builds on Luft and Ingham's (1955) Johari window but continues to be refined with labels changing accordingly. Despite its weaknesses, the beauty of this model is its ease of understanding for entry-level students of complexity theory. Moving beyond the most simple visual form, the layers of sophistication provide greater fidelity in addressing each quadrant (Figure 2.6).

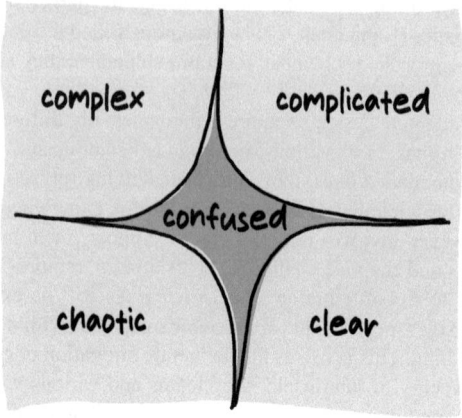

Figure 2.6 Cynefin Framework

One metaphor that helps with military audiences in particular is the symbolic "running to the sound of the guns" journey. Militaries have a Pavlovian response in moving rapidly from the *chaotic* quadrant using a clockwise direction to restore order in the *clear* quadrant (previously known as *simple*). A more sophisticated version of this concept is presented in Figure 2.7, where the vexing area of *chaos* is expanded based on extensions to the ontological and epistemological axes. Regardless of the model version – and somewhat counterintuitively – to dissolve complex systems into a more favourable state, designers must swim against the tide by deliberately moving in a counterclockwise direction to reach apparent chaos where cognitive biases (faulty mental models) are dissolved away, and ingenious options emerge. Although chaos is an uncomfortable place for military purists, this is precisely where divergent thinking flourishes the most.

2.4.1 OODA Loop

The OODA loop, published by military strategist and US Air Force Colonel John Boyd in 2018, is a decision-making framework that stands for "observe, orient, decide, and act." This model is designed to describe the cycle of decision-making in a competitive environment and is particularly applicable to military strategy. The first step, *observe*, involves gathering information from the environment. *Orient* refers to analysing this information and using it to update your current reality, taking into account new data, cultural traditions, genetic heritage, and personal experiences. *Decide* is the process of determining a course of action based on the orientation. Finally, *act* involves implementing the decision. The key to the OODA loop's effectiveness is speed

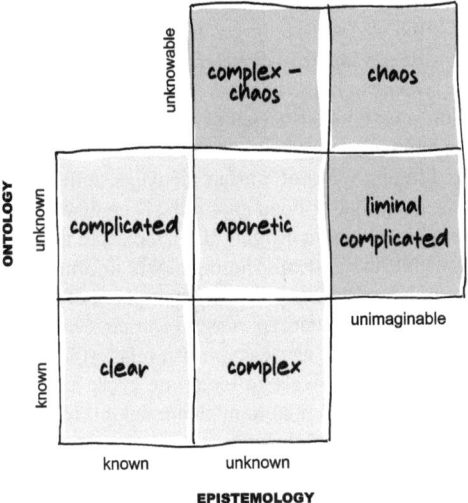

Figure 2.7 Uncertainty Matrix (Snowden, 2024)

and adaptability; the faster and more accurately one can move through these steps, the more likely they are to outpace and outmanoeuvre an opponent, making it a vital concept in both military and non-military strategic planning (Figure 2.8).

2.4.2 Model Integration

As previously stated, in the different types of warfare domains military planners often navigate, there's a need to shift their approach to manage the unpredictable nature of these environments effectively. This is where the integration of the Cynefin framework (Snowden & Boone, 2007) and John Boyd's

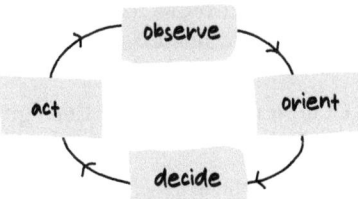

Figure 2.8 OODA Loop Framework

(2018) OODA loop becomes invaluable. The OODA loop underscores the importance of rapid observation, orientation, decision-making, and action to maintain a strategic advantage over adversaries and adapt swiftly to changing circumstances. Meanwhile, the Cynefin framework assists in navigating the problem classification domain. Together, they synergise to enable a dynamic and responsive approach to military planning and operations.

Within the different types of warfare domains, military decision-makers must embrace a sense-and-respond approach. They must conduct small-scale experiments or probes to gain insights and understand the patterns and interdependencies within the system. These insights inform subsequent actions, allowing for iterative adjustments and the emergence of new strategies.

The chaotic domain presents an even greater challenge, as it is characterised by the absence of clear cause-and-effect relationships. In this quadrant, commanders must act decisively to break the cycle of chaos and restore a semblance of order. It is essential to maintain flexibility, continuously assess the effects of actions, and rapidly adjust course as necessary.

The Cynefin framework, when applied in military contexts, provides commanders and planners with a structured approach to decision-making in complex and chaotic environments. By understanding the characteristics of each domain and employing appropriate methodologies, military organisations can enhance their agility, adaptability, and effectiveness in the face of uncertainty.

For the military, embracing the Cynefin framework and integrating the OODA loop can bring several benefits. It enables a more nuanced understanding of the operational environment, allowing for better-informed decision-making. By recognising the unique challenges posed by complex and chaotic domains, defence can develop strategies and tactics that are tailored to the specific characteristics of each domain.

Furthermore, the framework encourages a culture of agility and adaptability within defence. It promotes continuous learning, experimentation, and the ability to respond to changing circumstances quickly. This flexibility is particularly crucial in modern warfare, where the nature of conflicts and adversaries continues to evolve.

The Cynefin framework, along with the integration of the OODA loop, provides militaries with a basic decision-making framework suited for complex and chaotic environments. By leveraging this approach, militaries can enhance their operational effectiveness, optimise resource allocation, and navigate the complexities of modern warfare with greater confidence.

2.5   Exploiting Complexity Theory for Emerging Environments

The Cynefin framework is helpful in dissolving complex and chaotic military situations that militaries might encounter today or in the future for several compelling reasons:

Sense-making and contextual understanding: Complex and chaotic military scenarios often involve intricate dynamics and considerable uncertainty. The Cynefin framework provides a structured approach for comprehending and making sense of these complex situations. It enables military personnel to categorise and contextualise the nature of the challenges they face, facilitating a deeper understanding of the intricate context in which they operate.

Decision-making and strategy development: The Cynefin framework offers a valuable tool for decision-makers to assess the nature of a situation and determine suitable courses of action. By classifying the situation within the Cynefin domains (clear, complicated, complex, chaotic, and confused), military leaders can align their decision-making strategies with the specific context they are confronted with. This aids in selecting appropriate strategies and responses to address the complexities inherent in complex and chaotic military operations.

Adaptability and agility: Complex and chaotic military situations demand adaptability and agility in response to rapidly changing circumstances. The Cynefin framework emphasises the importance of continuous observation, experimentation, and adaptation. It encourages military personnel to be flexible, innovative, and responsive to emerging threats and opportunities, enabling them to adjust strategies and actions effectively.

Collaboration and interdisciplinary approaches: The Cynefin framework promotes collaboration and interdisciplinary approaches to problem-solving in complex and chaotic environments. These situations often require the integration of diverse expertise and perspectives. By leveraging the Cynefin framework, military teams can better understand the interdependencies and interactions between different elements, fostering collaboration and harnessing the collective knowledge and skills of diverse stakeholders.

Risk mitigation and failure avoidance: Complex and chaotic military environments inherently entail risks, and failure can have significant consequences. The Cynefin framework provides a structured approach to decision-making and action, helping to mitigate risks. It encourages military personnel to embrace uncertainty, anticipate emergent behaviours, and make informed choices, thus reducing the likelihood of costly failures and enabling more effective responses.

The Cynefin framework helps in complex and chaotic military situations as it facilitates sense-making, guides decision-making, promotes adaptability and agility, fosters collaboration and interdisciplinary approaches, and helps mitigate risks. By utilising this framework, militaries can enhance their capacity to navigate the complexities and uncertainties of modern warfare and effectively respond to evolving threats and challenges they might encounter in the present or future.

Senior military leaders can also exploit complexity theory to inflict greater chaos and confusion on an adversary. While the obvious role is to reduce complexity and chaos, the same principles can be used to give military commanders a transient advantage over the operating environment. This is obviously attractive in combat operations, but it can be equally useful in maintaining the initiative over looters and criminal gangs in post-disaster situations where there is a breakdown in law and order.

The creation of chaos in an adversary is done through the coordination and synchronisation of military forces using personnel, technology, and processes. The optimum blend of all three generates not just a symbiotic but also a synergistic effect known as superadditivity (Page, 2007). Asymmetry or force difference over the adversary will create localised chaotic domains. A key strength in this activity is establishing intelligence dominance through the exploitation of exquisite technologies. The impact this mass effect creates in an environment deliberately forces an adversary into chaos. This is not a new concept. Asymmetry or relative superiority has always been the goal of militaries, but the use of military design thinking helps generate more opportunities for winning. The military designer should aim to optimise for asymmetry and difference more quickly than the adversary and then concentrate forces to deliver relative superiority.

Case Study 3: The *Cookie Effect*

The *cookie effect* describes the emergent problems caused when breaking up a system. Based on the visual imagery of a cookie being smashed into multiple small pieces (Figure 2.9), the resultant isolated closed systems become harder to comprehend relative to when they were a single system. For example, the forced disintegration of a monitored terrorist network requires a completely different intelligence collection plan after

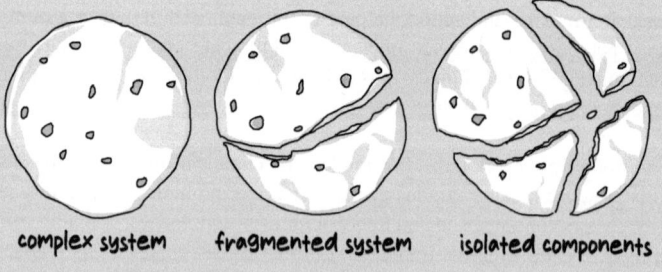

complex system   fragmented system   isolated components

Figure 2.9 The Cookie Effect

fragmentation. The cookie effect does not imply such action should never be attempted; it just means the consequences need to be carefully considered. In the case of an intelligence-gathering situation, a new collection plan needs to be devised and implemented prior to any controlled disruption action to avoid temporal loss of systems awareness. This example serves as a reminder for other complex situations where there is a temptation to assume the situation as only complicated prematurely and thus apply quick-fix resolution techniques.

3 Military Design Thinking

Framing rigidity means framing fragility.

3.1 Evolution of Military Design Thinking

The application of design thinking in military contexts – referred to as "military design thinking" – is seen as separate from civilian design thinking. In order to react to the unprecedented change defence forces are faced with today, a "military design movement" has emerged, which is experimenting with different ways of thinking, particularly the application of design thinking, to break open traditional military approaches to planning, problem-solving, and development (Zweibelson, 2017). This military design movement can be seen as returning to the art of warfare evident in Sun Tzu (544–496 BCE) and as retrieving the conception of war long overlooked in Carl von Clausewitz (1780–1831), whose text *On War* is foundational for many military practitioners. Sun Tzu is globally recognised for his profound understanding of the complexities of war and for teaching the importance of a rigorous process, which is fundamental to operational design.

A comparison has been drawn between Sun Tzu and Brigadier General (Ret.) Shimon Naveh's design approach. According to Mark Blomme (2015), Naveh's thinking resembles Sun Tzu's in its focus on out-thinking the enemy, exploiting surprise, and seeking asymmetric opportunities to undermine an enemy's strategy. Similarly, Clausewitz has been linked to military design thinking, although his conceptualisation of war has often been applied functionally by militaries. Clausewitz's military theory is generally seen as a broad framework for understanding all wars. For military design practitioners, however, Clausewitz's view of war as an instrument of policy is non-functionalist, emphasising that military actions must be understood within their political context, conceptualising war as an open system. Philipe Dufort (2017) offers a different perspective, identifying in Clausewitz a legacy of instrumental reflexivity that promotes overcoming cultural, ideological, or doctrinal certainties, fostering a more innovative approach to warfare.

DOI: 10.4324/9781003502180-3
This chapter has been made available under a CC-BY-NC-ND 4.0 license.

While these early military thinkers are worth noting, this chapter will focus on the recent formalisation of military design thinking, specifically the military design thinking that finds its origin in the work of Shimon Naveh. In February 1995, Naveh, who had just completed his PhD on the evolution of operational art at King's College London, returned to Israel, where he began working with a number of ex-brigadier generals on alternative manoeuvring concepts and reading non-military philosophical texts with the intention of improving the operational art of the Israel Defense Forces. Naveh's think tank was upgraded a few years later and became the Operational Theory Research Institute. There, Naveh developed a new planning methodology for Israel Defense Forces' generals he called systemic operational design (SOD) that drew upon systems theory, Soviet operational art, postmodern philosophy, and the practices of architectural design. However, the difficulty of the philosophy and concepts involved in SOD bred an opposition to the method in a number of generals who had completed Naveh's Advanced Operational Command Course. In 2006 two months prior to the outbreak of the Second Lebanon War, the Operational Theory Research Institute was shut down by the new Chief General of Staff, Lieutenant General Halutz. Halutz subsequently blamed SOD for the Israel Defense Forces' failure in 2006 and insisted that its 2004 planning manual *Operational Concept*, which contained a chapter on design and command, be abandoned. This is unfortunate, as it reflects a common tendency to blame the design methodology itself rather than considering other crucial factors such as the facilitators, the constraints imposed by leadership, the tactics, the strategic influences and the levels of design education and exposure. These variables are vital to any design capability (Mosely, Markauskaite & Wrigley, 2021). Consequently, this incident has unfairly tarnished the reputation of SOD on the international stage.

Naveh's work had already gained attention elsewhere. Faculty at the School of Advanced Military Studies in the United States (US) had developed an informal academic relationship with Naveh around the time of the founding of the Operational Theory Research Institute and taught their first course on SOD in 2005. At the time, US Army Training and Doctrine Command had also developed an interest in SOD, as it had become apparent that the war in Iraq was not a normal war and that it was necessary for the US Army to engage with culture. Students and faculty at the School of Advanced Military Studies continued to work on SOD and drew attention to the power of the methodology through their development of alternative approaches in the yearly war game Unified Quest. Design soon became a feature of US Army doctrine. It was included in the 2006 text *FM 3–24 Counterinsurgency*, the 2008 *US Army Training and Doctrine Command Pamphlet 525–5–500*, and by 2010, a chapter was devoted to "design methodology" in *FM 5–0* (Graicer, 2017).

The recent conceptualisation of military design thinking by Wrigley et al. (2021) bridges the gap between civilian design thinking and military operations. It recognises that conventional problem definitions often lead to predictable and conventional solutions. In contrast, military design thinking aims to break away from this paradigm by adopting an investigative and probing approach to problem-solving. Instead of immediately reacting to symptoms, the focus is on identifying the root cause and asking more interesting questions. Of particular note is the greater emphasis on divergent thinking and problem framing which is often overlooked in traditional reductionist military planning.

3.2   Defining Military Design Thinking

Recent developments in military design have seen its integration into most Western military doctrine publications and military education syllabi. Additionally, there is a growing number of military professionals who are now developing expertise in this field, contributing individually as military thinkers. This evolution has brought a blend of military and civilian design approaches, further enriching the field. Military design thinking's journey towards recognition as a distinct field of inquiry highlights its progression and growing significance. While it has lagged behind civilian design thinking in gaining formal acknowledgement, recent developments affirm its established place in both military strategy and academia.

Within the scholarly literature, there are very few cases where a definition of military design thinking is set out in an explicit manner. Philipe Beaulieu-Brossard and Philipe Dufort (2017) state that military design thinking means "the capability to understand a current conflict environment from a holistic perspective, to imagine a desired post-conflict environment and to realize it with counter-intuitive military and non-military means" (p. 2). Diren Valayden (2020) defines military design thinking as fostering "problem-solving capability in its end users" (p. 168).

The term 'design' is inherently multi-disciplinary, leading to various interpretations within the realm of innovation, which complicates any attempt at a universal definition (Wrigley, 2017). As a result, the literature lacks a general consensus or singular definition. Nevertheless, the authors of this book, drawing on their collective expertise in the field, present an enhanced definition that captures the full potential of military design thinking.

Military design thinking is a nonlinear cognitive approach to creatively address the dynamic challenges of increasingly complex military activities. It incorporates the key features of both systemic and product design methods but is tailored to the unique responsibility of promoting global peace and security.

While many commercial ventures might share some attributes, the scale and existential risk associated with defending a nation's national interests is incomparable. Military design thinking, therefore, is about maintaining a transient advantage over potential adversaries who pose a threat to our way of life. Military design thinking is applicable across the spectrum of competition and transcends all levels, from National Military Strategy to developing and integrating exquisite technological capabilities. After all, the crucial link of relevance must connect even the lowest levels of new equipment to government expectations. Military design thinking encapsulates the following elements that:

- Permit creative solutions to emerge by framing problems from diverse perspectives and magnitudes. If everyone arrives at the same solution state, then so too will the adversary, making diversity in thought essential.
- Addresses the challenges of modern warfare by fostering and supporting innovation across the entire spectrum of military challenges (eg. organisational culture to warfighting).
- Is agnostic of rank and domain hierarchies and applicable across all warfighting domains (land, sea, air, space, and cyber).
- Is interdisciplinary and relies on a range of cognitive reasoning tools, both individual and group.
- Employs a nonlinear approach to addressing the multifaceted challenges faced by the military, allowing for flexible and adaptive strategies.
- Encourages reflexive practice, enabling military personnel to rethink and redefine their understanding of complex adaptive systems.
- Allows for a holistic understanding of the current conflict environment and the capability to imagine and work towards a desired future state.
- Is characterised by its human-centered process where the impossible meets necessity, creating solutions that do not yet exist and enabling military organisations to gain relevance and advantage in emerging complex systems.
- Necessitates cognitive diversity to provide multiple perspectives on problems, freeing individuals from preexisting military mental models and encouraging critical movement between cognitive frames.
- Helps individuals move past traditional military mental models, fostering a more adaptive and innovative mindset suited to modern warfare's dynamic nature.
- Encourages a culture of experimentation and optimism, fostering an environment where creative ideas can flourish and innovative solutions to complex systems are actively pursued, moving beyond critical and reductionist thinking.

3.3 Military Design Thinking Characteristics

This section explores the distinctive traits that define military design thinking, offering a comprehensive breakdown of nine key categories as presented in Wrigley et al. (2021). Each category encapsulates fundamental characteristics essential for understanding the intricacies of military design thinking. Throughout this exploration, the aim is to define and elucidate each characteristic, offering insight into its significance within the context of military operations and strategic planning. By dissecting these traits, the nuanced approach that underpins military design thinking, highlighting its relevance and applicability in contemporary military contexts, is uncovered.

3.3.1   Complex and Adaptive Warfare

The idea that design thinking is the means through which to approach the new character of warfare in the 21st century is the most common characteristic that has been argued throughout the first two chapters of this book. The contemporary operating environment is seen as volatile, uncertain, complex, and ambiguous, and the traditional military decision-making process and joint operations planning process are understood to be inadequate for developing approaches to the counterinsurgency missions and asymmetrical conflicts that define this environment.

The operational methods introduced in the US directly prior to the development of military design thinking, such as effects-based planning and system-of-systems analysis, are also perceived to be ill-suited to the contemporary character of war. These older planning processes are recognised as mechanistic and reductive and, therefore, unable to provide interventions for complex challenges in which the parts are interrelated, dynamic, and interactive and cannot be separated from their environment. For problems of this kind, the analytical deconstruction of systems into their components cannot lead to effective solutions. Design thinking, however, has long been understood as a means of tackling wicked systems – contexts that are ill-defined and complex *(Buchanan, 1992)* – and it is for precisely this reason that armed forces around the world have integrated design thinking into doctrine.

Such complex and adaptive warfare is inherently dynamic, constantly evolving, and responsive to interactions within the system. Any modification to one part of the system can have ripple effects on other parts, underscoring the need for a comprehensive approach. Addressing these challenges in operations requires a complex systems approach, one that continuously reassesses and adapts to the ever-changing environment. This approach ensures that strategies remain relevant and effective in the face of future challenges.

3.3.2   Creative Agency

Military design thinking is a process characterised by creativity, producing innovative and novel interventions to complex adaptive systems. Creativity is widely regarded as an essential component of design in military contexts. Creative thinking is necessary for design to foster innovation, as it capitalises on imagination, insight, and novel ideas (Cardon & Leonard, 2010). This form of thinking leads to new insights, fresh perspectives, and inventive ways of understanding and conceiving systems (Ancker & Flynn, 2010). Although the terms are often used interchangeably, there is a need to distinguish between creativity and innovation. Creativity involves adapting knowledge gained from past experiences for use in new contexts, whereas innovation entails developing completely new approaches. This distinction underscores the value of both creative and innovative thinking in the military design process.

Desmond Stewart (2019) highlights the use of SOD by the Israel Defense Forces during Operation Defensive Shield in Nablus in 2002 as an exemplary case of harnessing creative agency. Soldiers innovatively searched buildings, not by entering through doors but by using explosives to create access points in walls. Stewart considers this an application of disruptive creativity to the interpretation of physical space, enabling the Israel Defense Force to change the rules of engagement successfully. This example illustrates how creative thinking in military design can lead to novel and effective operational methods, reinforcing the importance of creativity in fostering innovative solutions within complex adaptive environments.

The example of the US Special Operations Forces (SOF) further illustrates the significance of creative agency in military contexts. After the terrorist attacks on 11 September 2001, the involvement of SOF in the Iraq War was significantly supported by a substantial infusion of resources. This influx, however, led to a loss of the creativity and innovation that had previously distinguished SOF (Black et al., 2018). This stemmed from an overreliance on resources and a focus on hyper-conventional operational proficiency, which stifled creative thinking and adaptability. As a result, the organisation became constrained by conventional processes, narrowing its mission scope and diminishing the creative spirit that originally set SOF apart. To address this, Black and colleagues (2018) proposed that SOCOM adopt their methodology, known as the Special Operations Command design way, to reintroduce creative and innovative thinking.

3.3.3 *Problem Framing*

In military operations, problem framing is crucial for developing effective strategies. Problem framing describes the need to frame (define) problems (described in Chapter 2 as systems) and develop a standpoint or position that determines how a problematic situation is to be approached. This process involves clearly defining the problem, understanding the operational context, and establishing a standpoint that guides the approach to resolving the issue. It requires a deep understanding of the environment, recognising relevant stakeholders, and anticipating potential challenges. Incorporating creative and critical thinking allows military personnel to consider unconventional solutions and challenge existing assumptions. Problem framing is iterative, enabling solutions to be adapted as new information emerges and situations evolve. This comprehensive approach ensures that strategies are well-informed, adaptable, and capable of addressing complex challenges effectively, whether in counterinsurgency, cybersecurity, or humanitarian missions.

Mark Newdigate (2014), for example, writes that over time the evolving problem frame of US foreign policy objectives in Colombia has shifted and changed "from training para-military groups to counter left-wing guerrillas, to focusing efforts on counter-narcotics production and trafficking, to the establishment and support of a Colombian Joint Special Operations Command" (p. 18). Initially, this reflected the geopolitical tensions of the era and the desire to combat perceived threats from communist movements in Latin America. This initial approach represented a direct response to the insurgent activities that threatened the stability of the Colombian government and, by extension, US interests in the region. As the situation in Colombia evolved, so did the focus of US foreign policy. The shift towards counter-narcotics production and trafficking emerged as drug cartels gained prominence, creating new security threats and socio-economic challenges. This new problem frame indicates a recognition of the multifaceted nature of conflict, where the intertwining issues of guerrilla warfare and narcotics trafficking necessitated a broader shift in strategy that addressed not only military concerns but also the underlying socio-economic factors contributing to instability.

In this way, military design thinking is closely aligned with what constitutes a core aspect of design thinking practised in civilian contexts in which framing is central to design. Yet it must also be said that framing in the context of military design thinking is used, at least by those discussing the Army design methodology, in a modified sense. When Wass de Czege (2009) describes how to approach a problem using design, for example, the term *frame* functions as something like a synonym for *understanding* – that is, the development of the system and operating frames is the development of an understanding of the system in which the military will intervene and an understanding of what kinds of operations are possible within a system. These two framing activities subsequently lead directly to a clear problem statement, and hence the definition and framing of a problem, but framing is also used in a broader sense by Wass de Czege.

3.3.4   *Iterative Engagement with the Operating Environment*

Fundamentally design is an iterative process, where the problem and solution coevolve, is particularly evident in design systems, which are often not well-defined. This nonlinear process stands in contrast to the traditional step-by-step approach of military planning and is crucial for developing solutions to complex, ill-structured problems. Traditional sequential problem-solving methods are inadequate for addressing wicked problems. Instead, an iterative effort that initially focuses on framing the problem is necessary for effective operational design (Schnaubelt, 2009). Wicked systems are not well-defined or static and fail to account for complexity, evolving nature, and multiple stakeholders, requiring adaptive, iterative approaches that embrace uncertainty and interconnected factors.

Design facilitates not only iterative learning but also proactive, continuous engagement with the operating environment (Mosely, Wright & Wrigley, 2018). This approach allows planners to probe and gauge the system's response, much like threat actors do, making design, planning, and execution simultaneous and ongoing activities (Frumin et al., 2018). As planners analyse actors and their interactions, they adjust the design spaces within their plans and those of other actors. These adjustments occur because all actors are interconnected, causing cascading effects throughout the system. This iterative process helps planners identify and address the unintended consequences of their actions (Stofka, 2010).

3.3.5   Collective Force Team Collaboration

Design as a collective and collaborative process is essential in military design thinking, emphasising the importance of group discussions and teamwork for a comprehensive understanding of the problem environment. This approach leverages diverse perspectives, fostering a shared understanding among team members and enhancing creativity through the exchange of ideas.

Collaborative efforts facilitate critical feedback, allowing team members to challenge assumptions and refine solutions. Additionally, the iterative nature of teamwork supports adaptive learning and continuous improvement, ensuring strategies remain relevant and effective. By promoting synergy and cohesion, collaborative design processes improve decision-making and execution, ultimately leading to more successful military operations. It is the process by which military professionals arrive at a better, shared understanding of an environment, a problem, and a proposed solution (Proctor, 2011).

Collaboration is also understood to enable and inspire creativity and innovation. Collaboration and knowledge sharing, combined with creativity, are vital components of innovation. Understanding the egalitarian nature of design thinking, which embraces the sharing and acceptance of ideas, is essential for fostering innovation, with team diversity identified as a crucial factor in enhancing creativity. Findings suggest that teams composed of individuals with diverse backgrounds, credentials, expertise, roles, genders, personality traits, and problem-solving approaches tend to be more successful. Such diverse teams are perceived to engage in higher-quality discourse and develop more innovative solutions compared to homogenous teams (Perez, 2011). Within a military context, this approach seeks to leverage diverse perspectives across different services and disciplines, enhancing strategic decision-making and fostering more adaptive, innovative solutions to complex challenges.

3.3.6   Conception of the Environment as a System

Systems thinking and the use of diagrams and other visualisations to represent systems is a fundamental aspect of military design thinking that closely aligns with civilian design practices. In military contexts, systems thinking involves developing a holistic view of the operational environment, recognising the interdependencies and interactions between various components within the system. This approach allows for a comprehensive understanding of the complexities and dynamics at play, enabling military planners to identify leverage points and anticipate the ripple effects of their actions. By visualising the environment as a system, planners can better conceptualise relationships, feedback loops, and emergent behaviours, facilitating more informed decision-making and strategic planning. This systemic perspective is crucial for addressing complex adaptive challenges as it moves beyond linear reductionist approaches to embrace the interconnected nature of modern military operations.

Systems thinking encapsulates the critical and creative thinkers intending to explore change (Kulzy, 2019). Ervin Laszlo (1996) defines systems thinking as that which "gives us a holistic perspective for viewing the world around us, and seeing ourselves in the world" (p. 16). More specifically, systems thinking is necessary because, as Wass de Czege (2009) writes, "current mission environments present complex rather than complicated systems" (p. 8 – their parts are interrelated, dynamic, and interactive and cannot be separated from their environment). The conception of the environment as a complex system is therefore needed to assist military designers in understanding the relationships among the actors and elements of the environments with which they are involved. As Alex Ryan (2011) puts it, operations design – of which military design is his chosen example – is an inherently iterative and highly reversible form of design in which an ongoing relationship between the designer and the affected social group is established. Military operational designers are essentially part of the social system they are designing and, therefore, must conceive of their environment as open to flows of energy, matter, and information (Ryan, 2011). Today this is known as wicked systems.

3.3.7   *Critical Multi-Actor Analysis*

Critical multi-actor analysis is used to understand and manage the complex interactions between multiple actors in a given system. It involves analysing the roles, behaviours, and influences of various stakeholders to comprehend their collective impact on strategic scenarios better. This approach is particularly relevant in military and security contexts where interactions among diverse actors, such as different military units, allied forces, and civilian entities, must be coordinated effectively. It emphasises the importance of considering the dynamic and interdependent nature of these relationships. Using systems thinking and incorporating multi-criteria analysis helps in developing comprehensive strategies that account for the diverse motivations and actions of all involved parties.

In the context of critical multi-actor analysis, appreciating the values, perceptions, and biases of ourselves as well as our allies and adversaries is crucial.

This involves employing critical thinking to evaluate competing explanations of events to ensure that hypotheses are weighted according to evidence and to assess second- and third-order effects (Banach & Ryan, 2009). Cardon and Leonard (2010) emphasise that critical thinking entails asking appropriate questions, gathering relevant information, deriving sound conclusions, and effectively communicating those conclusions. These characteristics leverage critical thinking skills to distil vast amounts of information, identifying the most relevant elements for strategic decision-making. While some believe critical thinking opposes creativity, it is vital for reflexive practitioners of military design thinking, where it examines preconceived notions and fosters innovative strategies.

3.3.8   *Operational Approach Representation*

Operational approach representation, often referred to as a "design concept," is composed of narratives, graphics, products, systems, services, or a combination of these elements. These representations serve to provide order to complex situations and encapsulate them effectively. Narratives help articulate the underlying rationale, strategic goals, and envisioned outcomes, making complex scenarios understandable. Graphics, on the other hand, offer visual clarity by illustrating relationships, processes, and structures within the operational environment. Together, these elements facilitate a comprehensive understanding and communication of the strategic and operational approaches, enabling better planning and decision-making in military contexts.

Design concepts are a result of the design process, functioning as guidance for those responsible for the detailed planning that determines what actions are necessary to fulfil the proposed course of action (Banach, 2009). These concepts play a pivotal role in acting as the central idea that guides the overall direction, ensuring consistency throughout its development. In the US Army's School of Advanced Military Studies course on design thinking, students undertake a number of readings that focus on how a well-constructed narrative helps explain an unfamiliar or complex situation. Narratives, as one way to represent a design concept, are used during the design process to describe the problem to be approached, the environment in which the problem is positioned, and the proposed solution.

Visual graphics are similarly used to help make sense of the frames and to encapsulate the course of a proposed action. For example, graphics formed part of a developed design concept that visually represented the goal of a planned intervention in the post-2003 Iraq War. These visuals helped break down complex strategic objectives into more digestible components, facilitating clearer communication among diverse teams and ensuring alignment on the intervention's goals and execution. A detailed analysis of how visualisation techniques can be used to communicate complex ideas to command in a way that is straightforward and yet not reductive was generated by Moten et al. (2016). They offer this technique as a way to inform key decisions based on the complexity of these decisions.

Within the characteristic of visualisation, the concept of *narrative* found in military design thinking differs from civilian design thinking *storytelling* in that storytelling is, for the most part, utilised to understand the circumstances of those for whom a product, service, or experience is being designed. It does reflect, however, the realistic and strategic mixed-media narrative techniques described by Price et al. (2018), where they reported the demonstrated ability to sustain organisational innovation and frame new possibilities using such techniques.

3.3.9   Civilian-Centred

Design thinking is user- and human-centred, involving direct empathy and engagement with those impacted by a design, which is fundamental in civilian contexts. Empathy, a core aspect of design thinking, is crucial in defining its approach. While the perspective of those affected by a military design is considered, empathy in this context is practised indirectly through reflection on the cultural beliefs and values of the end users.

Empathy is employed in order to understand not only those who will be affected by military action but also members of the service who make the decisions (Collins & Mills, 2019). However, considering how a human-centred approach might also be applicable to defence and security challenges, such as pre-crisis stabilisation efforts, suggests that the defence and security communities could learn from the way the aid and development community is employing design (De Spiegeleire et al., 2014).

The problems caused by secondary empathy (through third parties) are evident, as the design team of the Proud American Battalion, in their deployment to Iraq in May 2009, found out. They had gathered information on the local populace from many sources before deployment, but upon arriving in Iraq, it became evident that their approach to the situation was not working. By reflecting on the interactions with community leaders that had occurred after they arrived in Iraq, it subsequently became apparent that the manner in which the environment and problem had been framed was incorrect. Missing from the design team's prior understanding of the situation was the community's fear of a resurgence of the (new) Ba'ath party (Proctor, 2011).

Furthermore, Proctor (2011) reports that although the fear these leaders held had been noted, it had been dismissed as white noise because it did not relate to the questions members of the battalion were asking. It was only once they empathised with this perspective that the battalion was able to find a successful approach. Designers need to recognise that there is no such thing as white noise when designing, rather than suggesting they attempt to understand the perspective of those who will be affected by the design.

3.4 Conceptualisations of Military Design Thinking

In order to build on these military perspectives, it is vital to understand the current conceptualisations of military design thinking. Design thinking within the military context is represented in different ways for different means and for different ends. Some understand military design thinking, either explicitly or implicitly, to be continuous with previously employed planning processes; others understand it to involve a radical rethinking of how the military approaches improving situations. As we have already mentioned, however, it is primarily used in the context of operations. Jackson (2019) offers a history of military design thinking and an account of its relation to civilian design thinking, describing the existence of two camps of military design thinking proponents. There are the purists who view military design thinking as a complex, interdisciplinary methodology that necessitates military personnel to reframe their understanding of a situation by questioning their core beliefs, thereby leading to innovative and adaptive solutions. On the other hand, there are the pragmatists responsible for integrating design into doctrine, who aim to make design thinking as straightforward and accessible as possible (Jackson, 2019).

Wrigley et al. (2021) propose that military design thinking is conceptualised in two main ways. These two approaches are described in the following section: a pragmatic approach that largely seeks to modify existing military operational art in a minimal manner, and a more reflexive practice that seeks to break free from traditional military modes of thinking and develop innovative approaches to the problems of the contemporary operating environment.

3.4.1 A Pragmatic Approach to Military Complexity

The pragmatic approach to military design thinking can be explained through the following case study. During the surge that occurred in Iraq in 2007 and 2008, there was tension between foreign al-Qaeda forces and Indigenous Sunni actors fighting against or resisting the coalition. Without design thinking, the solution to this problem would have involved a critical vulnerability analysis of the al-Qaeda and Sunni forces and then attacks on these groups. By using design, however, a different approach was developed. Commanders and staff exploited the tension between al-Qaeda and the Sunni forces "and achieved an improved state of affairs in which coalition troopers and Iraqi Sunnis were pointing their rifles not at each other, but toward [al-Qaeda] fighters" (Perez, 2011, p. 48).

This is surely an improved outcome, but it does not involve reframing ontological commitments. The strategy employed here is grounded in a pragmatic understanding of leveraging existing tensions to achieve a favourable tactical position. Indeed, the notion that "the enemy of my enemy is my friend" is a well-established principle in military and diplomatic strategy, having driven the collaboration between the Western Allies and the Soviet Union in World War II. This principle operates on the basis of temporary alliances formed against a common adversary without necessitating a fundamental change in the underlying perceptions and identities of the involved parties.

This approach, while effective in the short term, does not address the deeper, more systemic issues that fuel conflict. By not challenging or reframing the ontological commitments – the core beliefs and assumptions about identity, allegiance, and enmity – such strategies may fall short of fostering lasting peace and stability. For instance, in the context of Iraq, the coalition's tactical manoeuvre to pit al-Qaeda against Sunni forces was a clever exploitation of immediate circumstances, but it did not resolve the underlying socio-political grievances and power dynamics that contributed to the conflict.

Reframing ontological commitments would involve a deeper, more transformative process. It would require addressing the root causes of conflict, such as sectarian divisions, political disenfranchisement, and economic disparities. This could involve initiatives aimed at building inclusive governance structures, promoting economic development, and fostering a shared sense of national identity that transcends sectarian lines. In contrast to the exploitation of existing tensions, a design thinking approach that seeks to reframe ontological commitments would focus on long-term solutions and sustainable peace. This would entail engaging with local communities to understand their perspectives and needs, co-creating solutions that address their core concerns, and fostering an environment where former adversaries can reimagine their relationships and roles within a shared framework of peace and cooperation.

Over the years, many have argued for the introduction of a "define" step at the beginning of the military decision-making process (Cooney, 2012). Cooney (2012) provides an example of the benefit of this approach. A platoon leader who receives a mission to secure a bridge can use the "five whys" design tool to determine that achieving the mission will require preventing enemy forces from utilising the bridge, a task best achieved using the hilltop overlooking the bridge. The platoon leader might then define the problem as the defence of the hilltop so that the bridge can be secured. Cooney does not suggest how the platoon leader would have approached this mission without defining the problem, but it is clear here that military design thinking is not conceived as a process leading to triple-loop learning. This deeper topic exploration involves extensive research to uncover the root causes of issues, ultimately leading to more appropriate decision-making (Tosey et al., 2012). Such understanding underpins the principles of design thinking.

The design methodology known as SOD has been labelled by some as asking practitioners to cultivate a habit of integrating reason and intuition to make decisions so that later decisions might be made more rapidly (Mazzara, 2011). This understanding is also evident in Perez's (2011) definition of design as something that "does nothing more than give a bit of structure to those periodic conversations any commander has with his staff officers to improve his appreciation of the mission" (p. 43). If design thinking is understood merely as something enacted so that operational decisions can subsequently be made more quickly or just a formalisation of the processes already employed by commanders, then it cannot be understood to be aimed at innovation or as a unique methodology that provides new ways to approach the complex character of warfare in the 21st century. This pragmatic approach does not allow for the full advantages that design thinking can offer the military. Perhaps, as seen with SOD, external variables influenced the perceived limitations of the pragmatic approach. Therefore, a more thorough and balanced evaluation of this approach should be conducted before condemning an entire field of work.

This pragmatic approach to design thinking boils down to thinking critically but in such a way that institutional norms are not a barrier to new ideas. It seeks to incorporate innovative thinking within the existing frameworks, making it practical and applicable without causing disruptive upheaval. However, there are inherent challenges in balancing innovation with adherence to established protocols.

It is clear that unless there is a concerted effort to change the underlying paradigm through which the military views the world, the true potential of design thinking for the military will not be achieved. This requires a shift from merely embedding design thinking into doctrine as a set of tools or processes to fundamentally transforming the way military personnel approach problem-solving and strategic planning.

A pragmatic approach, though valuable, must not become a compromise that stifles the full spectrum of innovation. It should instead serve as a bridge, facilitating a gradual yet profound shift in military culture. This includes fostering an environment where critical thinking is encouraged and radical ideas are not just tolerated but actively explored.

For design thinking to reach its true potential within the military, it is essential to cultivate an openness to reframing problems, questioning core beliefs, and embracing interdisciplinary methodologies. This transformation would enable military personnel to develop adaptive solutions that are not only effective in the short term but also sustainable in the long term.

The pragmatist position in military design thinking is far more prevalent (Wrigley et al., 2020), amongst the academic and military community and also (it seems) in practice than the reflexive position (Figure 3.1). This is likely due to the pragmatic nature of the US Army design methodology as it is presented in doctrinal texts such as *FM 5–0* and *ATP 5–0.1*, which are widely available. Reflexive practitioners of military design thinking, on the other hand, largely reject the codification of military design thinking into doctrine. The fact that reflexive military design thinking practitioners identify a

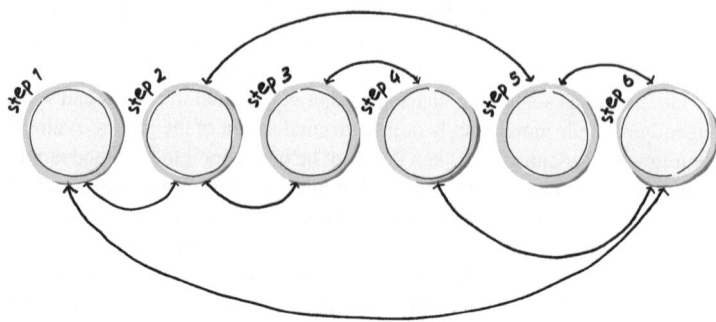

Figure 3.1 Pragmatic Approach to Military Design Thinking

fundamental discrepancy between their approach and doctrine is not surprising. It is a discrepancy that reflects the well-documented incongruity observed in the civilian practice of design between the iterative and cyclical nature of design thinking and the linear processes that already exist in the (business) organisations where design thinking is introduced.

3.4.2   A Reflexive Approach to Military Complexity

Today's military personnel contend not only with traditional physical threats but also with sophisticated information and cyber weapons wielded by adversaries. This dynamic and multifaceted battleground demands innovative approaches to planning and execution. Traditional command-and-control hierarchies, with their emphasis on obedience, are ill-suited to navigate these emerging chaotic environments. The fusion of technology and complexity necessitates characteristics beyond those historically required at strategic command levels. Reductionist methodologies, once relied upon to simplify and characterise warfare, are now challenged by the saturation of technology, overwhelming the cognitive capacities of planning staff.

Military design thinking advocates for a departure from reductionism and embracing complexity, which is a methodology that goes beyond traditional frameworks. It champions a probing and acting approach in chaotic environments, facilitating sense-making and adaptive planning. Unlike conventional methods, military design thinking allows for continual adjustments to mission statements, acknowledging the fluid nature of the environment and problem statements. This flexibility enables design teams to adapt their frameworks based on evolving environmental understanding.

To become proficient in military design thinking, personnel must undergo structured learning experiences in controlled environments under the mentorship of experienced professionals. With time and experience, the process

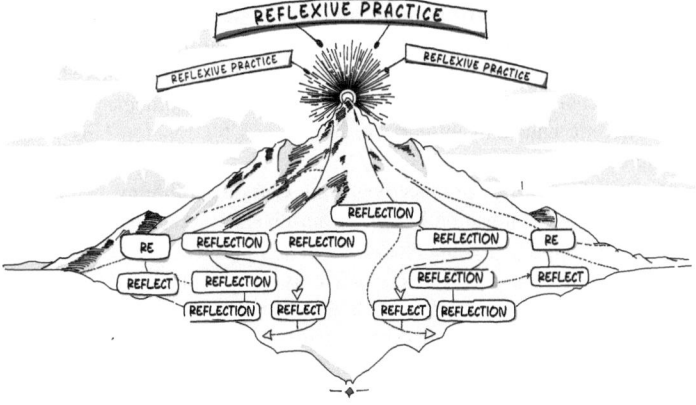

Figure 3.2 Reflexive Approach to Military Design Thinking

becomes instinctive and reflexive, empowering practitioners to craft adaptive responses to complex challenges.

By embracing a reflexive and innovative approach (Figure 3.2), military organisations can effectively navigate the complexities of the contemporary operating environment, enhancing their operational effectiveness and adapting to the ever-changing nature of warfare. Design plays a pivotal role in this reflexive approach, acting as a catalyst in the process. Reflexive innovation involves continuous self-assessment, learning from experiences, adaptability through prototyping, inclusivity of multiple perspectives, ethical considerations, collaboration and dialogue, emphasis on process over product, and responsible innovation. These reflexive characteristics ensure military strategies remain effective, ethical, and responsive in complex operating environments. These are explained further here:

Continuous self-assessment and system reframing: The reflexive approach to innovation, characterised by continuous self-assessment, aligns with the military design principle of reframing problems. This involves not only assessing current practices but also rethinking the entire concept of military engagement, which includes challenging traditional thought processes and assumptions.

Learning from experience and user-centric approach: Reflexive innovation emphasises learning from both successes and failures, which mirrors the user-centric approach in military design. This involves understanding defence challenges from the perspective of those on the ground and adapting strategies based on real-world feedback and experiences.

Adaptability through prototyping and iteration: A reflexive approach requires adaptability, a trait shared with the military design principle of prototyping and iterative development. In military contexts, this involves continually testing, learning, and adapting strategies based on the dynamic nature of military operations and the relationships within complex systems.

Inclusivity of multiple perspectives and cross-disciplinary collaboration: Reflexive innovation benefits from diverse viewpoints, much like the cross-disciplinary collaboration essential in military design. Bringing together varied expertise and perspectives is key to challenging established norms and exploring comprehensive ideas in military strategy and tactics.

Ethical considerations and visceral cognitive processes: Ethical considerations in reflexive innovation align with the use of visceral cognitive processing and communication in military design. By making complex concepts tangible and understandable, military design can ensure innovations are ethically sound and consider the broader impact, including social and environmental effects.

Design dialogue and systems thinking: The collaborative and dialogic nature of reflexive innovation is echoed in the systems thinking approach of military design. Understanding the broader context and interdependencies in military operations ensures that new strategies are well-integrated, responsive, and considerate of various systemic impacts.

Emphasis on process over product innovation: Reflexive innovation values the innovation process as much as the outcome. This approach ensures that military strategies are not only effective but also considerate of learning implications and long-term dissemination.

Case Study 4: Pick a War . . . Any War

Complex adaptive systems appear in every military operation. Recent conflicts in Ukraine, Gaza, and Afghanistan highlight that the dangers of treating such environments as merely complicated – when supposed solutions can be imposed through force domination – are fraught with disaster. The age-old whack-a-mole mentality of annihilation might win the war, but does not win the peace. For example, President Putin's attempt to subjugate Ukraine before it joined NATO backfired when Finland and Sweden promptly joined because of his invasion. However, Sweden's membership was not a complicated matter – it was complex. Türkiye and, in turn, Hungary both seized the opportunity to veto Sweden's application until other demands were met. The final resolution required multiple interventions from multiple nations where concessions were granted.

At a more tactical level, right-wing political influences in Israel sought to use heavy-handed military responses on Palestinians in Gaza following the 7 October 2023 attacks. While such a reductionist military response might have offered certain short-term successes previously, the fact that the region has been an intermittent war zone since the dawn of history suggests that traditional military thinking might not actually be the answer. Israel's actions in Gaza led to unprecedented reactions around the world, with global protests and diaspora feeling even more unsafe. Military action on the ground meant Israel exhausted an unmeasurable amount of international goodwill from previously friendly or even neutral countries.

Addressing the challenges of a complex system superficially is akin to providing temporary pain relief without considering long-term implications. Military design thinking necessitates a holistic consideration of the entire system and the exploration of the multi-order effects of actions through various epistemic perspectives. Effectively tackling complex systems requires an understanding of root causes and the implementation of changes that influence key stakeholders. Critical nodes analysis is one of many tools available to systems thinkers to facilitate this process.

Design thinking is about going beyond the obvious. The old adage "when all you have is a hammer, everything looks like a nail" speaks to the futility of forcing an incorrect solution on a misunderstood problem. Thus, military design thinking is not just about having the right hammer in your toolbox, nor even if another tool in the toolbox would be better – it is about asking if maybe a completely new tool needs to be developed.

4 Nurturing Creative Mindsets

For the strength of the Pack is the Wolf, and the strength of the Wolf is the Pack.
— Rudyard Kipling, *The Jungle Book* (1894)

4.1 Building an Innovative Culture

Promoting a stronger culture of innovation is a complex challenge. Unlike a simple time-speed-distance problem, with its reductionist, formulaic way of deriving a single right answer, culture is a diachronic system (see Chapter 2) requiring enduring leadership. Furthermore, influencing wicked systems to a more favourable state cannot be *solved* by pulling a single lever – it requires multiple concurrent interventions (Ackoff et al., 2006). Complex challenges demand complex responses. Fortunately, innovation has and continues to occur throughout Western militaries. The purpose of this chapter, therefore, is to examine the positive influences on innovation with the goal of protecting and promoting them. Being a complex system, though, means the topic needs to be examined through multiple lenses.

Interventions can be broadly clustered into macro- and micro-level initiatives (Heltberg, 2022). To explore existing and possible emerging opportunities, this chapter approaches the issue from both directions. It begins by considering individual excellence at the micro level before considering organisational culture – or macro level. Although both are important, they each approach the issue from different directions and eventually overlap in the middle (Figure 4.1).

4.2 The Micro-Level Lens: Individuals

Individuals are the building blocks of any organisation. Some would argue they are your greatest capability. Typically, these people represent a spectrum of cognitive abilities and personalities. Militaries are no different, with their spread of personnel representing a sizeable cross-section of society. There are those who are quite happy coming to work and being told what to do; they

DOI: 10.4324/9781003502180-4
This chapter has been made available under a CC-BY-NC-ND 4.0 license.

top-down strategic initiatives

the nexus

bOttom-up inspired individuals

Figure 4.1 Promoting a Culture of Innovation

do not aspire to be decision-makers. At the other end, there are those who are naturally curious and enjoy the thrill of being challenged. In between are those who have ideas to contribute but need encouragement to speak up. This section begins by considering the intellectual geniuses before questioning how to coax the middle percentile out of their mediocrity.

4.2.1 Individual Geniuses

Some people appear to be inherently talented in seeing what others cannot. These naturally gifted individuals have an uncanny ability to think differently about problems and foresee opportunities. Despite the best efforts of psychometric recruiting tests (screening for pattern-recognition intelligence over nonlinear creativity) and the ongoing self-stabilising influences for organisational conformity, these ingenious thinkers can be real assets when properly understood and managed by command. Not only can their cognitive skills be harnessed for designing innovative technologies to bridge capability gaps, but they are also valuable for contestability during systemic design activities. In popular parlance, these valuable assets are known as *mavericks*.

4.2.2   Understanding Maverickism

Maverickism is often used to describe those who operate on the limits of the permitted. To many, they are seen as boundary pushers who challenge conventional wisdom and seek innovative ways to improve the world. To others, however, mavericks are nonconformists who disrupt the comfortable stability of ordered systems. In both cases, progress relies on these people, but their professional acceptance in hierarchical organisations can be problematic when not appreciated.

The modern interpretation of a maverick remains largely true to its origins. The term was named after Samuel A. Maverick, a 19th-century American rancher and politician who was renowned for the unconventional approach of not branding his cattle (Maverick & Maverick, 1921). The real reason, however, was that Samuel and his wife Mary moved off their farm in 1854, leaving their unfortunate slave, Jack, unable to control the vast wandering herd on his own (Maverick, 1942). Neighbours finding unbranded calves within their herds began referring to them as "Maverick's cattle" (Figure 4.2), with the name eventually shortening and then spreading to anyone demonstrating originality and unfettered spirits (Thomas, 2001).

As a psychosocial personality type, maverickism has a modest following in the literature. Although widely used as an adjectival moniker and occasionally appearing in print (Sibley, 1973; Swallow, 1959), it was in publications

Figure 4.2 Maverick Cattle Are Not Afraid to Occasionally Distance Themselves From the Herd in Search of Improved Methods

such as *The Maverick Executive* (McMurry, 1974) that the positive benefits of innovation began to be appreciated. More recently, empirical studies have attempted to explore this phenomenon. A study in 2012, for example, found a correlation of extraversion, openness to experience, and low agreeableness with those identified as mavericks (Gardiner & Jackson, 2012), and related studies into innovators found linkages to the attributes of unstructured thinkers, neurodivergence, and intuition (Marrin, 2007). Similarly, others have explored the positive benefits of having mavericks in the military (Gladwell, 2021; Mandel, 2020; Matheson, 2007; Teichert, 2023). A series of publications from Ree Jordan have raised the topic's profile even further (Jordan, 2019, 2022s, 2022b; Jordan et al., 2021, 2023). And while there are a number of defining characteristics, as seen in Table 4.1, perhaps the most important distinction is that mavericks are altruistic in their actions. Mavericks do not break rules for personal gain.

Table 4.1 Maverick Characteristics

Recognising military mavericks	
Passionate	Mavericks are driven by a deep passion for their work and a strong commitment to making a meaningful impact within defence. Their enthusiasm and dedication fuel their relentless pursuit of positive change.
Change advocates	Mavericks actively advocate for change and improvement. They question established practices, challenge the status quo, and champion innovative approaches that can drive transformative outcomes within the organisation.
Rank bypassers	Unconstrained by hierarchical structures, mavericks are unafraid to challenge traditional authority and bypass rank when necessary. They prioritise the value of ideas and outcomes over strict adherence to chains of command.
Fearless	Mavericks exhibit fearlessness in the face of challenges. They are unafraid to voice their opinions, take calculated risks, and push boundaries to drive positive change within defence.
Disruptive	Mavericks have a natural inclination for disruption. They actively seek opportunities to introduce new ideas, methodologies, and approaches that challenge conventional thinking and pave the way for innovative solutions.
Inquisitive	Mavericks are curious and intellectually courageous. In seeking deeper insights and understanding, they ask tough questions so as to uncover underlying issues and identify new possibilities for improvement.
Permission vs. forgiveness	Mavericks embody the principle of "asking for forgiveness rather than permission." They are action-oriented and empowered to take the initiative, make decisions, and take responsibility for their actions to drive progress.

(*Continued*)

Table 4.1 (Continued)

Recognising military mavericks	
Innovative thinkers	Mavericks are characterised by their ability to think innovatively. They possess a knack for generating novel ideas, thinking outside the box, and proposing creative solutions to complex problems.
Risk-takers	Mavericks embrace calculated risks. They are willing to step into uncharted territory, experiment with new approaches, and venture beyond their comfort zones to drive positive change and pursue organisational success.
Open to new ideas	Mavericks maintain an open-mindedness to new ideas and perspectives. They actively seek out diverse viewpoints, embrace novelty, and cultivate a culture of creativity and innovation within defence.
System experience	Many mavericks have valuable experience within the military, providing them with a deep understanding of its intricacies, limitations, and potential for improvement. This firsthand knowledge enables them to navigate the organisation effectively and to propose informed solutions.
Early adopters	Mavericks are quick to embrace new technologies, methodologies, and practices. They eagerly adopt emerging tools and approaches that have the potential to enhance operational effectiveness and drive positive change.
Confidence	Mavericks exude confidence in their abilities and ideas. They believe in their vision and can effectively articulate their viewpoints, inspiring others and rallying support for their initiatives.
Relentless	Mavericks display a relentless drive and determination in pursuing their objectives. They do not easily give up in the face of challenges and setbacks but persistently work towards their goals with unwavering commitment.
Rule challengers	Mavericks challenge unnecessary rules and regulations that hinder progress and innovation. They prioritise practicality and common sense over bureaucratic constraints, advocating for streamlined processes that enable efficient and effective decision-making.
Forward thinkers	Mavericks possess a forward-thinking mindset, anticipating future challenges and opportunities. They proactively seek out ways to adapt and stay ahead of the curve, driving continuous improvement and long-term success.
Willingness to try	Mavericks are not afraid to try new things. They embrace experimentation, recognising that failure is a stepping stone to success and an opportunity for valuable learning and growth.
Challenger mentality	Mavericks inherently possess a challenger mentality. They actively challenge existing norms, assumptions, and limitations, seeking to overcome barriers and achieve breakthrough results.
Rank diversity	Mavericks can be found at any rank within defence. Their influence is not limited to a particular hierarchical level, as their impact and effectiveness stem from their ideas, actions, and ability to inspire others.

(Continued)

Table 4.1 (Continued)

Recognising military mavericks	
Fearlessness of failure	Mavericks are unafraid of failure. They understand that taking risks and exploring new possibilities inherently involve the potential for setbacks. However, they embrace failure as a valuable learning opportunity and remain resilient in the face of adversity.
Outcome-focused	Mavericks prioritise training and capability outcomes above all else. They are driven by the desire to improve the military's operational effectiveness, ensuring that the organisation is prepared to fulfil its mission successfully.
Encouragers	Mavericks are focused not only on their own success but also on empowering others. They actively encourage and inspire their colleagues, fostering a collaborative and supportive environment that champions innovation and growth.
Strong presence	Mavericks have a strong presence and are not afraid to stand tall in their convictions. They exhibit confidence, charisma, and a commanding presence that can influence and inspire those around them.
Community of networks	Mavericks often form networks and communities within defence. These networks serve as platforms for collaboration, idea sharing, and mutual support, further amplifying their collective impact.
Intelligent	Mavericks demonstrate a high level of intelligence and intellectual capability. They possess the cognitive agility to grasp complex concepts, analyse information critically, and propose innovative solutions.
Charisma	Mavericks are often charismatic and have an innate ability to connect with and influence others. Their compelling personality and persuasive communication skills enable them to inspire and rally support for their ideas and initiatives.
Caring	Mavericks genuinely care about the organisation and its mission. Their passion and commitment extend beyond personal ambition as they strive to make a positive difference and contribute to the greater good of the military.
Resilient	Mavericks exhibit resilience and strong-mindedness in the face of challenges and opposition. They remain steadfast in their pursuit of positive change, undeterred by setbacks or resistance.
Trust in people	Mavericks place a high level of trust in individuals rather than relying solely on bureaucratic processes. By believing in the capabilities and potential of their colleagues, they foster a culture of empowerment, collaboration, and accountability.

What sets mavericks apart is their ability to approach things differently, defying bureaucracy and traditional military policy norms. Their success stems from their authenticity and their inclination to create something unique rather than conforming to expectations. This concept is also known in the design field as a design innovation catalyst. Despite differing origins, similar attitudes, actions, and principles apply (Wrigley, 2016). However, within most militaries, mavericks are often seen as troublemakers, disrupting the conservative status quo. Ironically, it is these very behaviours that enable them to achieve breakthrough ideas in the first place. These risk-taking, rule-breaking individuals possess characteristics, qualities, skills, and behaviours that should be revered within the military. Unfortunately, these valuable assets are often underappreciated during peacetime due to the organisation's conservative nature. Ultimately, though, embracing mavericism would enable militaries to out-think their adversaries. The process of acculturating innovation, however, is long and slow. Expecting innovation to blossom suddenly just because a great power conflict occurs is somewhat naive especially when all the mavericks have left the building.

4.2.3   *Encouraging Introverted Geniuses*

Militaries have a number of intellectual geniuses who shun the spotlight. These are the people who often hold higher-level degrees, such as doctorates or research masters, but might be less well endowed with the social mastery to communicate their ideas. For example, it is not uncommon to encounter intelligence analysts who are brilliant at their "dots-on-maps" job but have no aspirations to lead others. While the military often values their day job contributions, there is scope for these people to be employed as team members in cross-disciplinary situations. Similar to mavericks, these are the perceptive ones who can serve as contestability agents in workshops outside their own subject expertise. Giving them a voice can be very similar to the way middle-percentile personnel are empowered.

Some intellectual giants are not attracted to the military lifestyle. The current approach employed by most Western militaries is to simply employ such talent in roles where a uniform is not required. Well-known entities such as the Defense Advanced Research Projects Agency (DARPA) in the US are good examples of specialist STEM experts who tinker away in their laboratories, making rockets go faster or miniaturising exploding pens. To be fair, not all projects are product-focused, but the majority of civilian innovators seem to be employed for technological innovation. The problem for militaries, though, is that this add-on approach does not necessarily translate to boosting the cognitive agility of uniformed systems thinkers.

To enhance the military's innovative mindset, they need personnel with a strong understanding of national security concepts, which will allow their self-initiated epiphanies to be nested within a national security framework. Those who enhance exquisite technologies are great and should be supported, but more needs to be done. Perhaps some additional military secondments into the laboratories might enhance innovative mindsets. Ultimately, however, it is the global context that drives concepts and concepts that drive capability requirements. Exquisite technologies are fantastic, but only if they are in response to the contextual demand from the high levels, not the driver of them (Figure 4.3).

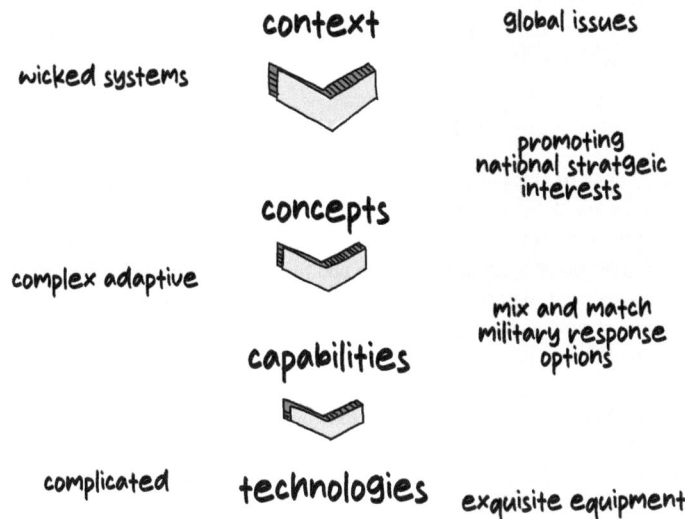

Figure 4.3 Military Hierarchy of Needs

4.2.4   *Empowering the Middle Percentile*

While the majority of defence personnel do not fall into the highest echelons of geniuses, they still have much to offer. The most obvious group is those with subject-matter expertise (SME) – although, as was discussed in Chapter 3, these personnel are not always the best to have in the room when contesting current processes. However, those with minimal SME bring another dimension that is, in fact, vital to challenging the status quo.

Systemic design thinking requires epistemically diverse members within a collaborative project. Due to the enormous attributes of a complex adaptive – or wicked – system, a single ingenious thinker is unlikely to bring sufficient diversity to the challenge. Even if they are familiar with all aspects, they are likely to suffer from the SME biases outlined previously. Thus, to maximise the degree of understanding in a wicked conundrum, multiple representatives from across a spectrum of impacted disciplines are needed. Under the careful leadership of an astute facilitator, every person will get a voice.

Epistemic diversity does not just favour the gifted and talented; rather, epistemic diversity is essential for any member of defence. Those who bring a rich diversity are those with non-traditional backgrounds. Those who come from reservist units and have a different civilian job are particularly valuable, as are those who transfer laterally from other services, militaries, or even roles within their own service. Many Western militaries are increasing their diversity of new recruits as well. While this can help better represent the society they protect, it has the added benefit of diversifying innovative ideas. The current shift to recruiting more culturally and linguistically diverse members is definitely something to celebrate.

4.2.5   *Summary of Micro-Level Considerations*

Mavericks are altruistic with their motives. Unlike those who simply break the rules for their own advantage, true mavericks challenge the status quo to improve the way the organisation achieves its goals. But, mavericks need to be understood by their leadership and managed accordingly. Furthermore, mavericks need to know when to push and when to conform. There is a time to laugh, a time to dance, and a time to play by the rules.

Defending a nation requires a champion team, not a team of champions. Having a number of isolated mavericks employed in an ad hoc manner across dispersed commands is never going to achieve full organisational excellence. Even when partially protected from institutional inertia by their local leaders, their limited critical mass will always struggle to achieve real progress. The challenge here is not just how to empower them with more freedom to shine but how to increase their numbers. Fortunately, there are ways to increase the pool of innovative talent – but it requires macro-level change from the organisation's most senior leaders.

Case Study 5: Desmond Doss

While mavericks are often associated with aviators and a famous Hollywood depiction of a certain US Navy pilot in particular, there are many others from across the services. Most Medal of Honour and Victoria Cross recipients, for example, become famous for not only their selfless acts of bravery but also boundary pushing to get the job done. In fact, the true story of Desmond Doss (Figure 4.4) typifies this very situation. Desmond Doss exemplifies a maverick in every sense. Despite enduring significant ridicule and attempts to push him out of the military, Doss challenged the norms and expectations of his organisation by steadfastly resisting the pressure to conform, particularly during his training and beyond. His refusal to use weapons, grounded in his pacifist beliefs, defied the military's traditional combat approach. Yet, his commitment to serving for truly altruistic reasons—his desire to save lives rather than take them—drove him to persist in the face of ongoing criticism. Even when faced with ridicule from peers and superiors, Doss went above and beyond, risking his own life to save 75 comrades during the Battle of Okinawa. His determination, bravery, and adherence to his convictions, despite all odds, exemplify the essence of a combat maverick: someone who defies convention and perseveres for the greater good.

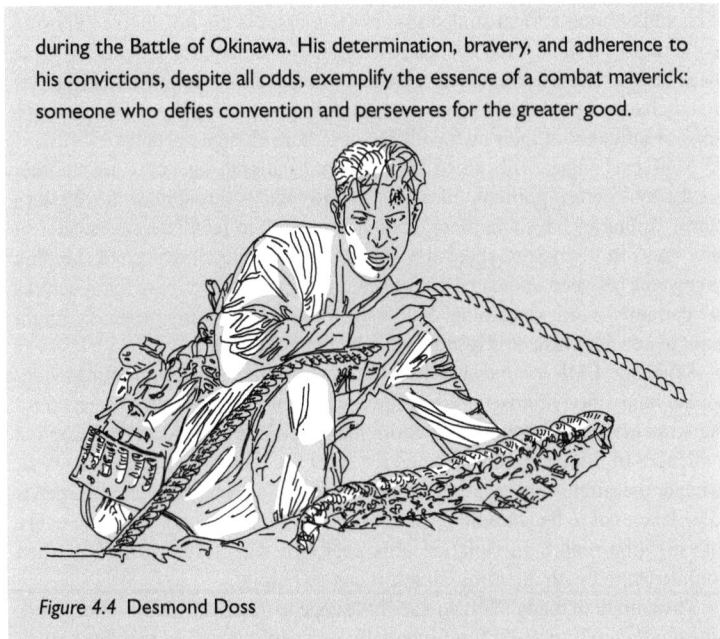

Figure 4.4 Desmond Doss

4.3 The Macro-Level Lens: Organisational Culture

4.3.1 Stewarding the Culture

This section explores opportunities to sustain and promote organisational culture change. It begins by considering the influential levers of empowering those in key units that exist to design and implement culture change. These include not just higher headquarters departments, with their organisational design psychologists, anthropologists, sociologists, and human geographers, but also local hubs and key leadership positions. It also extends to the schoolhouses that can influence the next generation of movers and shakers. The section on organisational culture then considers current approaches employed around the world for promoting greater workplace innovation. The process of actually leading organisational change is discussed in Chapter 5.

4.3.2 Targeting the Key Influencers

Throughout the Western world, there is much to learn from allied militaries about how they promote a culture of innovation. The longstanding practice of sending personnel to attend other nation's professional military education (PME) courses has always been a great success. An added bonus of exchanging international students on PME courses is the invaluable exposure to alternative thinking and epistemic diversity for both parties. Bringing these ideas back to a home country is a piece of the puzzle for improving organisational change – especially when the returning graduates are posted into positions of influence at their own colleges or culture change agencies.

Staff exchanges with allied and regional national agencies are another great way to cross-pollinate ideas. As already exists throughout the Western world, militaries have a number of exchange posts to facilitate the transfer of new ideas in a range of specialist areas. Those countries that have standing exchanges between dedicated culture change agencies and innovation centres are certainly worth sustaining. Similarly, those who are less proactive might want to consider how this resource can be better tapped.

Domestic PME courses can be a force multiplier when promoting widespread innovation mindsets. While the relative emphasis on alternative thinking strategies varies around the world, the schoolhouses tend to have captive audiences of those who are selected to "level up." Importantly, though, PME is never the single-shot solution to all the military's problems. Care still needs to be taken not to have training (or education) as the default solution to every investigation report. In isolation, this approach will never work, but when complemented with other initiatives, it can be helpful.

The culture of many PME courses has scope to mature. In some cases, modernising the culture away from internally competitive (e.g., a yearlong selection board) to something more akin to incentivising cognitive excellence over false metric achievement scores. The legacy system is often hindered by an instructional mindset rather than an open-minded liberal education. To succeed in post-course promotion and command appointments, some reporting systems are still focused on competitive achievement. This encourages students to shield their weaknesses and show off their strengths – a counterproductive reward system. Colleges that still make a quantitative score the determinant for ranking graduates can learn from those institutes that have embraced a more appropriate approach of rewarding those students who sought to work on their weaknesses and avoided choosing subject options or essay topics that they were already experts in. Thus, colleges that already incentivise actual professional development over year-long selection boards are encouraged to sustain this philosophy, while others might want to revisit their opportunities to improve.

Case Study 6: PME Course Reporting

The New Zealand Command and Staff College commissioned a study into their learning philosophy and the way their students were motivated to learn (Simons, 2009). A major finding of this investigation was that most students prioritised their quantifiable university grades over non-assessed military topics – despite the latter being more likely to make them better workplace officers. In fact, one student arrived at the course already an expert on terrorism and managed to complete the entire program without reading a single new book. He used his existing knowledge to complete assignments on leadership, strategic studies, operational studies, and the like, all with a terrorism flavour. In effect, he gained very little from the year-long course, and yet his final report was glowing. As a result of this major investigation, the New Zealand Defence College introduced a new learning philosophy and redesigned its end-of-course reports.

The college moved the numerical academic grades to the last page of the course report and refocused the first page on students' openness to self-improvement. It now reported on their courage to explore topics that were new and cognitively challenging rather than subjects that simply gave them a high achievement score. In short, this incentivised what PME is supposed to be about rather than the legacy hijacking by promotion boards, the irony being that the rankings such boards previously received were not actually highlighting the best people for promotion. Innovation comes from intellectual curiosity and risk-taking. Fundamentally, this is about learning *how to think, not what to think,* to facilitate transferrable cognitive agility skills (Figure 4.5).

Figure 4.5 Professional Military Education Learning

4.3.3   *Promoting Local Initiatives*

Beyond compulsory PME courses for career progression are some excellent short courses. Although these may not have the prestige of longer courses and sometimes fall into extrinsically motivated micro-credentialism, they are offered with the intent of *shifting the dial*. Whether a strategic initiative or simply supported at a local level, a number of workshops and projects invoke design thinking concepts that undoubtedly encourage participants to think differently about their everyday decision-making options. Such events are known variously as *design thinking, alternative thinking, quick wins*, and *adaptive red teaming*, among others (Kardos & Dexter, 2017). Regardless of their name or the style they employ, such events have the potential to be a greater encourager of innovative mindsets. Endorsing these activities is an organisational culture lever, yet the courses themselves focus on individual development, thus the nexus of top-down and bottom-up influences. A similar argument, however, could be made for the pop-up innovation hubs.

Innovation hubs on camps and bases were seen as a great way for senior leadership to endorse the need for greater workplace creativity (Figure 4.6). In many cases, however, such venues started with great fanfare but soon became dusty storage spaces. Although well intended, the rooms full of 3D printers and virtual reality goggles were targeting the tangible icons of innovation (products), not the process itself. Such resources are indeed supportive, but the

Figure 4.6 Innovation Hubs Are Often Launched With Great Fanfare, Yet Many Struggle to Sustain the Momentum

reality is that defence needs innovation to be endemic across the organisation, not cloistered away in a hobby shed. So, although the provision of innovation (maker space) hubs sends an encouraging strategic message, more needs to be done to integrate this with the workplace. One aspect of innovation hubs that is worth celebrating and promoting is the various workshops on design thinking (or whatever alternative label is used) for middle and senior leadership. Having inspired junior personnel at a hub is a start, but if it is not endorsed in the workplace, then the challenge is insurmountable. Perhaps the secret here is to incentivise more unit-level leaders to attend innovation workshops and learn what ideas their enthusiastic staff bring back. Getting quick *runs on the board* with real-world problems is in itself an enabler; it helps to chip away at the tomes of policy and doctrine that inhibit more abstract innovations.

Normalising military design thinking requires greater endorsement from senior leadership. Regardless of its actual name, the concept of empowering junior personnel to have a voice is what really counts. While the assimilation of alternative thinking strategies is progressing, its glacial rate may not be quick enough to sufficiently prepare Western militaries for the growing war clouds on the horizon. The benefits of military design thinking have already been highlighted in Chapter 3, and although various enlightened leaders have been willing to normalise this in the workplace, there is scope for greater encouragement by senior leadership. As shown in a recent European study, innovation hubs, workshops, and courses may not be enough on their own (Heltberg, 2022). Graduates can potentially act as "translators" back to the workplace, but the evidence is by no means certain.

Organisational culture is not a collection of isolated initiatives. As a wicked system, every lever pulled by senior leadership will influence how other dimensions impact the organisation's dynamics. While often, there is indeed a cohesive strategy for rolling out culture change initiatives, these are not always as integrated as they could be. In fact, there is a strong case to suggest this top-down approach itself needs disruption to help overcome the repeated failures to date (Herrero, 2008). Some more traditional thinkers remain either consciously, or subconsciously, of the belief that isolated levers can be pulled without impacting others. In STEM disciplines, this approach is known as "setting one to zero," and it works well for complicated problems where parts of a system are independent of others. In moving from *sustain* to *promote*, the importance of multiple concurrent initiatives needs to be thought through as much as possible during dissolving workshops.

4.3.4 *Structural Changes*

Incentivising innovation in key leadership roles is vital. While opportunities vary across militaries, common incentives include recognition and advancement based on the promotion of innovation cultures. One example of this is protecting commanding officers from punishment when members of their team try but fail to implement better ways of doing things. Those about to enter their dream role are thus conditioned into risk avoidance for their tenure. "Not on my watch" is a killer of innovation. The flip side is those in key roles are rewarded for trying new things.

Inspired leaders seek to promote innovation enablers. Sometimes, this includes not just promoting positive cultures but also managing those who vacillate on the fence. A host of innovation inhibitors, such as high operational tempo, understaffing, inadequate funding, and fear of reputational damage, are all examples of the reasons military leaders report frustration with innovation. Organisations that actively seek to offset such barriers are encouraged not just to sustain but also to promote their efforts. Furthermore, sharing their success stories with others not only transfers the learning but also encourages and inspires.

Organisations that build the promotion of innovation into leadership promotion criteria are role models to others, though some target only certain levels. For example, one particular investigation in Australia found that *promoting innovation* was an annual reporting attribute for all rank levels except for the most influential level of O5 (Young, 2017). Countries that reward innovation at all rank levels are to be applauded.

Publicly celebrating success is a popular tool for many who lament the demise of innovation. This obviously sends a message from the highest echelons and is thus a cultural enabler for lower-level commanders to emulate. An oft-discussed issue with innovation, though, is celebrating failure. For many, this remains a sensitive issue to celebrate publicly. One quirky initiative conceived at the Royal Australian Navy's innovation hub in Sydney was to award a prize to the teams who failed the most during an innovation competition. The need to normalise such courage is vital if it is to become a genuine cultural artefact.

Case Study 7: Playing the Joker

Maverickism and a bold sense of adventure are often associated with the joie de vivre of junior pilots on flying squadrons. Although this apparent oxymoronic risk-taking lifestyle seems at odds with the dangers of flying, there is a commensurate need to learn which boundaries are sacrosanct and which ones are fair game. For example, Tony Kern's discussion on "red rules" (written in blood) and "brown rules" (bureaucratic crap) is instructive here (Kern, 2009, p. 87).

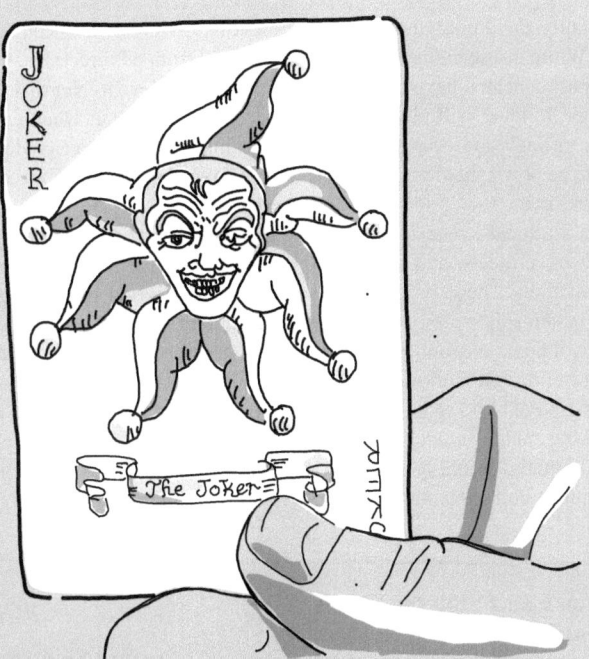

Figure 4.7 Joker Cards

To help sustain a culture of daring courage, it is not uncommon for commanding officers of flying squadrons to issue annual *joker cards* to their pilots (Figure 4.7). The bearers of such a "get-out-of-jail-free" card can then play it if something goes wrong. Commanding officers do, of course, retain a caveat for more serious line crossings, but in general, this system helps maintain a healthy workplace culture of boundary pushing.

An extension to this case study involves a former Chief of the Royal New Zealand Air Force, who took the joker card to a new level. Sensing a lack of innovation across the entire organisation, one of his first announcements was awarding an annual joker card to every single person in the air force. A bold move indeed, but it sent a very strong message to not just those under his command but also those watching on. Excessive bureaucracy had become rampant, and things needed to change. The days of so-called brown rules were over, and innovative mindsets were going to be celebrated.

4.3.5   *Summary of Organisational Levers*

Organisational culture is a top-down process that needs forward-leaning leadership. While in-use cultures can and do vacillate from all directions, those in the highest echelons have their hands on the biggest levers. Not that pulling on a specific lever will always yield the desired results, but when carefully thought through and properly implemented, many organisation-wide initiatives can be powerful in promoting a positive environment for a more innovative workforce.

Organisational levers such as strategic endorsement and tangible support for courses, projects, and structural processes play a crucial role in fostering innovation. These processes can include incentivised mechanisms and reward systems, like promotion criteria and public recognition. Without these supports, innovation cannot thrive in a stagnant culture. Facilitators are fundamental in design workshops, as they cognitively challenge participants to avoid cognitive blind spots within the group. A good design facilitator is vital in breaking mental models and encouraging cognitive friction, which promotes alternative thinking and innovative outcomes. However, mastering this skill requires both contextual knowledge and design expertise (Mosely et al., 2021).

4.4   You Can't Stop the Waves, but You Can Learn to Surf

Democratic nation militaries enjoy harvesting the fruits of innovation but seldom relish its essential ingredient of rule-breaking. Indeed, the very serious business of closing and engaging with an enemy demands a strong culture of discipline and obedience. However, this need must be balanced against the imperative for having a faster spinning OODA loop than the adversary. There is no glory in having the most disciplined warfighters if they are soon subjugated to the will of the victors.

A key message of this chapter is that culture is a complex adaptive system that requires constant feeding. No single-shot input will suffice, nor should influences always come from the same direction. Numerous macro-level conditions need to be constantly refined (such as incentivised promotion criteria or greater delegation of decision-making), while at the same time, micro-level empowerment of individuals also needs investment. Examples of these extend beyond just PME workshops to workplace tolerance and encouragement. Those who are naturally talented mavericks need special consideration, while the larger "silent majority" needs coaxing out of their shells through the empowerment of design thinking principles. Such ideas are the focus of the remaining chapters.

5 Military Organisational Constraints and Conditions

My adversary is my teacher, and my ego is my enemy.

5.1 Understanding to Overcome

Innovation within any organisation encounters a myriad of challenges stemming from various factors. Within the military landscape, this is even more acute. To fully grasp the conditions and constraints of the military innovation culture, this chapter dissects the array of challenges that hinder its progress. It explores how organisational culture, rigid hierarchical structures, and traditional operational mindsets collectively act as formidable barriers to innovative thinking. The chapter will also shine a light on the less visible but equally impactful psychological barriers, such as risk aversion and the fear of failure, which permeate the ranks and decision-making processes within the military.

In addressing these challenges, a blend of initiatives is proposed. These include fostering a culture of calculated risk-taking, promoting cross-functional collaboration, encouraging cultural leadership, and leveraging resources and infrastructure. Each of these potential strategies aims to transform a military innovation ecosystem into a more dynamic, responsive, and forward-thinking entity capable of adapting to the ever-changing landscape of modern warfare and security challenges.

As explored in the preceding chapter, individual innovators and their unique characteristics undoubtedly play a significant role in driving military innovation. However, militaries must go beyond focusing solely on the talents of select individuals and instead cultivate a robust innovation network – a dynamic environment that fosters creativity, collaboration, and experimentation across all levels of the organisation. By understanding these intricacies organisations can position themselves to navigate the complexities of the military landscape with agility and resilience.

DOI: 10.4324/9781003502180-5
This chapter has been made available under a CC-BY-NC-ND 4.0 license.

5.2   Resistance to Innovation

While innovation clearly requires inspired individuals who collaboratively miti-gate cognitive biases (Chapter 4), there is a macro-level dimension as well. Not every organisation wants to change its culture. As with various individual cogni-tive blind spots, there are evolutionary reasons why self-stabilising influences are fiercely protected by organisations. Despite the perceived threats to egos and power, change inevitably comes at a cost, even though innovation and change are mutually dependent. Furthermore, many decision-makers are simply unaware of the connection between protecting business-as-usual structures and the negative impact it has on organisational promotion of innovation.

Organisational fear of complexity theory is as prevalent in the military as it is in industry. To date, attempts to win over organisational culture have been largely limited to isolated pockets of mavericks struggling to inspire the organisation through a disjointed bottom-up campaign. Most senior leaders – those with the greatest ability to influence organisational culture change – appear trapped in the belief that innovation hubs with virtual reality helmets and 3D printers will somehow promote innovation. In their defence, such facilities do demonstrate a desire to promote innovation, but the hubs soon become dusty. Perhaps the two biggest causes for their demise are the fact that they focus often on individuals (not epistemically diverse collaborations), and they are too product-focused.

The military has a justifiable reputation for authoritarian discipline and conformity. Indeed, such attributes have served militaries well for millennia when confronting the chaos of war (Finkel, 2019). Yet the days of an army marching over the horizon to fight an isolated and dichotomous battle of good versus evil have given way to the complex globalisation of international rela-tions. Governments and non-state actors alike now employ a dynamic num-ber of influences to gain competitive advantage – the use of military force typically being one of last resort. And even when militaries *are* mobilised, they almost always operate in harmony with a sophisticated national or coali-tion strategy. Such orchestration involves a wide variety of influences such as economic, diplomatic, legal, political, and informational, among others, which are often reduced to simplified ontologies such as ASCOPE, PESTLE, PEMSII, DIME, or even DIME-FIL (Rodriguez et al., 2020).

To contribute meaningfully to the complexity of a nation's international relations policies, militaries of democratic countries must accept that they cannot simply enforce rigid doctrines on when or how to fight. Furthermore, rapid advances in technology and greater access to higher education mean that adversaries are more inclined to exploit ingenious innovation to outmanoeu-vre traditional doctrine. The days of deeply entrenched tactics, techniques, and procedures are declining.

At the tactical level, with inexperienced personnel in dangerous and chaotic situations, the need for discipline remains vital. However, as decision-makers gain experience and advance in responsibilities, they need to appreciate that complicated solutions are no longer effective when influencing complex adaptive systems. At the operational level and beyond, decision-makers need to transition to interdisciplinary (known as *inter-agency* but extends to the *whole of nation* and *non-governmental organisations*) collaboration for nuanced influences.

5.3 Organisational Constraints to Military Innovation

In the military context, the pursuit of innovation, creativity, and design is often challenged by a multitude of constraints deeply embedded within the organisational culture and structure. The blockers are often immediately listed – all the reasons why a new idea won't work or why it can't happen within the organisation. These constraints encompass various factors, including a pervasive fear of failure that discourages risk-taking and experimentation, rigid command structures that stifle autonomy and time constraints that prioritise immediate operational demands over strategic thinking. Additionally, resistance to unconventional ideas, a fear of criticism or ridicule, and an inherent aversion to change further impede the exploration of innovative solutions. Outdated infrastructure, inadequate resources, and restrictive policies pose significant barriers to implementing creative ideas and embracing new approaches.

Addressing these constraints requires a concerted effort to foster a culture that values creativity, encourages collaboration, and empowers individuals to challenge existing norms and pursue innovative solutions to complex challenges. Despite innovation occurring at all levels within the military, numerous influences impede certain larger initiatives. These constraints, described in Table 5.1, reflect the unique global military culture. Identifying these constraints is the first step towards addressing them systematically and breaking each down individually.

Table 5.1 Constraints on Military Innovation Culture

Military Innovation Constraints	
Organisational cognitive load management systems	**Rules and regulations** *The military emphasises adherence to rules and regulations and is often characterised by bureaucratic reductionism while simultaneously promoting doctrinal understanding and adherence to standard operating procedures (SOPs). This is understandable for safety on weapons ranges in order to foster instinctive reactions; however, this adherence becomes a constraint when it fails to cultivate personnel who understand when to deviate from established rules, thus empowering them to make informed decisions and adapt to dynamic situations* (DePaul, 2022; Ervin, 2020; Kern, 2011, p. 87; Jans & Frazer-Jans, 2004; Odell, 2022; Kern, 1999).

(Continued)

Table 5.1 (Continued)

Military Innovation Constraints

Predictability and conformity

In military operations, predictability is often valued for maintaining order and safety (particularly in peacetime), but adherence to rigid doctrine can make actions easily anticipated by adversaries, and it can stifle innovation and autonomy, thus hindering the ability to think independently and adapt to dynamic situations (Milevski, 2012; Orak & Walker, 2021).

Institutional inertia

Larger organisations are often characterised by slow decision-making processes, resistance to change, collective reluctance to adapt, and the difficulty of changing direction, whereas smaller militaries often find it easier to implement change and initiate new initiatives due to their agility and flexibility (Frank et al., 1996).

Hierarchical culture and disempowerment

The hierarchical culture of the military, driven by rank and authority, can impede innovation by hindering open communication and collaboration. This rigid structure often prioritises adherence to established practices, stifling the input and ideas of junior ranks. The entrenched culture of legacy fosters resistance to change due to a deep-rooted familiarity with traditional approaches. This resistance arises from a lack of exposure to new technologies and or a reluctance to venture beyond an individual's comfort zone (Ervin, 2020; Bell & Patterson, 2005; Schmidtchen, 2001).

Organisational structure and processes

When individuals are inundated with immediate tasks and operational requirements, they struggle to find the time for other innovative-related activities. The adherence to established practices represents a failure to adapt to evolving threats or technologies, and individuals risk overlooking opportunities for improvement and miss out on innovative solutions to emerging challenges. When individuals become entrenched in the system that governs their daily workload, it can be exhausting, and they may lose intrinsic motivation or incentives to innovate. This can result in complacency or disengagement, leading to a stagnation of ideas and a reluctance to explore new approaches (Tucker, 2002).

Proof of concept demands

Testing new concepts without guaranteed outcomes is often discouraged because of rigid demands for proof of concept and a lack of culture surrounding prototyping. This highlights the necessity of fostering a culture that allows for experimentation, seed funding, and support to learn from and iterate upon ideas without the need for full validation (Assink, 2006).

Linear problem-solving tools

Military innovation is impeded by the rushed application of linear problem-solving tools like IMAP, CMAP, and SOPs, as individuals often prioritise quick solutions over thorough analysis, neglecting the potential for more effective use of these tools (Harężlak & Rosa, 2019; Margetts, 2016; Zweibelson, 2023).

(Continued)

Table 5.1 (Continued)

Military Innovation Constraints

	Artificially bounding complex systems *People treat problems as complicated when they are really complex in nature; therefore, bounded rationality is used incorrectly* (Stewart, 2009; Rosser, 1999). **Micromanagement (technology-enabled communications)** *The proliferation of technology-enabled communication channels has facilitated remote control of field decisions from centralised bases, leading to disempowerment and fostering a culture of micromanagement in real-time combat situations* (Miller, 2012). **Deference for quantitative metrics** *The military's preference for quantitative metrics often leads to the unnecessary measurement of factors that don't truly reflect impact. Relying on quantifiable statistics for reassurance and reporting purposes sometimes results in inaccurate reporting and a focus on numbers rather than meaningful outcomes* (Greene, 2020).
Organisational general expectations (enforced)	**Legal, moral, and ethical behaviour** *While adhering to these principles is essential for maintaining integrity and trust as well as order and discipline within the military, this adherence can inhibit the exploration of innovative solutions due to strict compliance requirements. Enforced rules can stifle creativity and impede the exploration of unconventional ideas, hindering the military's ability to adapt and evolve effectively* (Jans & Cullens, 2010; Kilcullen et al., 2001; Pearce & Saul, 2008). **Short-term objectives** *Political agendas and short-term opportunities often exert significant influence on decision-making processes and the realisation of strategic visions within both holistic military approaches and individual service branches. For instance, large-scale acquisition projects and targets necessitate meticulous long-term planning and sustained resource allocation. While on paper, life-of-type studies are conducted, in practice, defence may procure sophisticated assets without adequately supporting them with essential maintenance and staff training, thereby jeopardising long-term operational capability. A preoccupation with short-term objectives or immediate operational exigencies, at the expense of longer-term strategic priorities, can constrain the attention and resources allocated to innovative endeavours. Achieving a harmonious equilibrium between short-term imperatives and long-term innovation demands strategic foresight, judicious resource management, and a concerted effort to cultivate innovative capabilities for the future. Moreover, the practice of annual budgetary cycles, where any underspend is often perceived as a failure, further exacerbates these challenges and hinders the pursuit of sustained innovation within military contexts* (Tucker, 2002).

(Continued)

Table 5.1 (Continued)

Military Innovation Constraints

	Unexamined traditions
	Unexamined traditions, often upheld with the justification of "we've always done it this way," can stifle innovation and hinder progress. It's important to question and reassess traditions when they are no longer appropriate, as demonstrated by such decisions as permitting personnel to grow a beard, to ensure openness to change and adaptation in pursuit of effectiveness and relevance (Abdukakharovna, 2020; Burk, 1999).
Organisational cultural reward mechanisms (controlled through reinforcement)	**Promotion and career management systems**
	Some militaries do not incentivise innovation. In certain systems, innovation is not directly linked to career advancement or promotion criteria. However, there has been a recognition of the need to revise these systems to encourage and reward innovation. International militaries should assess whether their promotion criteria support and incentivise innovation effectively. By aligning promotion and career management systems with innovation goals, militaries can foster a culture that encourages creativity and forward-thinking approaches to problem-solving (Solbach et al., 2022; Young, 2017).
	Spheres of influence, control, and concern
	When the three spheres of influence, control, and concern are out of proportion in an innovation idea, it can hinder progress. This imbalance may lead to individuals neglecting the important aspects necessary to advance their innovation initiatives. For innovation to thrive, individuals must be attentive to all relevant factors and actively address areas within their influence, control, and concern. Failure to do so can impede the successful implementation and development of innovative ideas (Covey, 1991).
	Mavericks/Positive deviants/Rule breakers/Functional/ Constructive deviants
	Individuals who challenge the status quo, known as mavericks, positive deviants, or rule breakers, often face resistance despite their potential to drive innovation. While some view them as catalysts for change and encourage their disruptive behaviour, others perceive them as obstacles to be punished. There exists a spectrum of mavericks, each viewed differently by the organisation, with some celebrated for their innovative thinking and others marginalised for their perceived disruptions. To foster a culture of innovation, it's essential for military organisations to recognise and respect the contributions of mavericks, encouraging their unconventional approaches while mitigating any negative impacts on cohesion or discipline (Jordan, 2022; Mandel, 2020; Chung & Moon, 2011).
	Pay for performance
	Pay for performance can inadvertently incentivise individuals to prioritise personal financial gain over the mission and goals of the organisation. When personnel are motivated solely by extrinsic rewards, such as increased pay packets, they may lack the intrinsic motivation needed to drive innovation. This "dash-for-cash" mentality can lead to a focus on self-interest rather than pursuing the best outcomes (Rynes et al., 2005).

(Continued)

Table 5.1 (Continued)

Military Innovation Constraints

Organisational cultural attributes (largely uncontrolled)	**Peter principle (promotion above capability)** *Individuals are promoted to their level of incompetence by placing them in positions where they lack the necessary skills or expertise to lead innovation efforts effectively. When personnel are promoted based solely on their past performance or seniority rather than their ability to drive innovation, it can result in ineffective leadership and a lack of vision for implementing innovative solutions. This can lead to missed opportunities for innovation and hinder the organisation's ability to adapt to changing circumstances or embrace new technologies and strategies* (Chamorro-Premuzic, 2013; Peter & Hull, 1969). **Bias for action** *In the context of running to the sound of the guns, a bias for action can hinder military innovation by fostering a preference for immediate, tangible results over more comprehensive, divergent thinking. This bias may lead to a focus on resolving urgent or critical incidents without sufficient consideration of alternative perspectives or innovative approaches, ultimately limiting the exploration of unconventional solutions and impeding the organisation's ability to adapt to complex challenges in the long term* (Wadham & Connor, 2023). **Frog in boiling water** *When junior personnel's ideas are underutilised and not respected or encouraged, this stifles the creativity and contributions of those who may offer fresh perspectives and innovative solutions.* (Murray & Millett, 1998; Zisk, 1993). **Eating their young (burning out innovative junior members)** *"Eating their young," a phenomenon where senior leaders harshly criticise and discourage innovative junior members due to their own challenging career paths, can stifle creativity and motivation within the organisation, hindering the development of fresh ideas and potential future leaders* (Noworol et al., 2017). **Resignation to status quo** *In being resigned to the status quo, individuals lack the motivation to challenge existing norms or invest energy in innovative endeavours. Without adequate incentives, such as recognition or rewards, there's little appeal in deviating from the established way of doing things. Additionally, the fear of risking other job responsibilities can discourage individuals from pursuing innovation, leading them to prioritise stability. This tendency towards self-stabilisation ultimately makes it more appealing for bosses to manage, perpetuating the cycle of maintaining the status quo rather than fostering innovation* (Kern, 2009).

(Continued)

Table 5.1 (Continued)

Military Innovation Constraints

Lack of collective passion

When there's a lack of shared enthusiasm or commitment to driving change and pursuing innovative solutions, individuals may be less inclined to invest time and effort into exploring new ideas or challenging existing practices. This can result in a culture of complacency, where innovation is deprioritised in favour of maintaining the status quo. Without a collective passion for innovation, the military may struggle to foster the creativity, collaboration, and forward-thinking mindset necessary to address evolving threats effectively and achieve strategic objectives (Sørensen, 1994).

Institutionalism vs. Occupationalism profession

Institutionalism fosters a genuine sense of belonging and loyalty but may also cultivate rigid adherence to established protocols and resistance to change, hindering the adoption of innovative practices and technologies. The compartmentalisation of expertise within occupational specialties can impede cross-disciplinary collaboration and the sharing of innovative ideas, further constraining the military's ability to adapt and innovate effectively (Moskos, 1977).

Expectation of positivism

By prioritising metrics and tangible outcomes, there is a tendency to favour known methodologies and solutions, stifling the exploration of unconventional ideas or approaches that may not fit within traditional measurement frameworks. This reliance on positivist approaches limits the military's ability to embrace ambiguity, creativity, and adaptability, hindering innovation efforts aimed at addressing complex and evolving challenges (Reed et al., 2004).

Last person standing theory

The last person standing theory is where innovators leave due to frustration with a lack of support or resistance to change. Those that remain may perpetuate the status quo, becoming resistant to new ideas or initiatives. This leads to a depletion of innovative talent and a reinforcement of conservative attitudes, impeding the military's ability to adapt, evolve, and effectively address emerging challenges (Johansen et al., 2013).

Change fatigue

Change fatigue can hinder innovation by diminishing motivation, focus, and resources as individuals become overwhelmed by continuous organisational changes (Bergkvist & Karlsson, 2019).

Nepotistic cronyism

Influential service personnel in secret societies or fraternities can act as blockers to innovation by favouring members within their group, potentially stifling new ideas that originate outside of the established network (Barker, 2014; Ellis-Smith, 2018).

(Continued)

Table 5.1 (Continued)

Military Innovation Constraints

Conservativism

While conservatism is often embraced to maintain stability and mitigate risks associated with rapid change, excessive adherence to traditional practices and a reluctance to embrace new ideas can inhibit innovation. By prioritising stability over adaptability, missed opportunities occur. Striking the right balance between conservatism and innovation is essential to ensure the military remains agile, responsive, and capable of addressing evolving threats and challenges (Finkel, 2019).

Iron colonels

Iron colonels is a fictional concept that describes a group of particularly strict or authoritarian colonels within a military hierarchy. Alternatively, it might be a term coined within a specific organisation or culture to describe colonels known for their unyielding adherence to rules or protocols (Kalms, 2020).

Frozen middle

The term frozen middle *refers to a phenomenon where there is a lack of movement or progress in the mid-level ranks of the hierarchy. These mid-level leaders, often referred to as the "middle management," become "frozen" in their ways, as the term suggests, and resistant to change. This stagnation hinders organisational agility and innovation as new ideas and initiatives struggle to gain traction within this segment of the hierarchy. Overcoming the frozen middle often requires deliberate efforts to engage and empower mid-level leaders, encourage a culture of innovation and openness to change, and provide the necessary support and resources for implementing new initiatives* (Jackson & Humble, 1994; Williamson, 2023; Jensen, 1998; Duguid & Goncalo, 2015).

Risk aversion

A risk-averse culture within the military can deter individuals from taking the necessary risks associated with innovation. This aversion to risk stems from a lack of trust between senior officers and junior personnel, inhibiting open dialogue and collaboration. Often, the fear of failure, rejection, and loss of control discourages individuals from taking risks or proposing innovative ideas (Jungdahl & Macdonald, 2015).

Organisational training (convergence) over education (divergence)

Professional military education

Professional military education (PME) can become a blocker to innovation when it is delivered solely as training rather than true education. Although the concept of PME is valuable, its effectiveness is compromised when institutions prioritise rote memorisation and adherence to historical case studies over critical thinking and forward-looking analysis. By focusing solely on "what to think" instead of "how to think," PME fails to equip military personnel with the skills and mindset needed to anticipate and address future challenges effectively. This approach limits innovation by fostering a culture of conformity rather than encouraging creative problem-solving and adaptability (Simons, 2009; Antrobus & West, 2022).

(Continued)

Table 5.1 (Continued)

Military Innovation Constraints

Formal training

This approach emphasises a one-size-fits-all methodology, prioritising conformity and obedience over creative problem-solving. By instilling a culture of adherence to predefined procedures, formal training limits individuals' ability to think innovatively and adapt to dynamic situations. This focus on conformity not only stifles creativity but also perpetuates a culture where innovation is marginalised, hindering the military's ability to respond effectively to evolving challenges (Grant, 2013; Evans, 2008; Palazzo, 2012).

Do schools kill creativity?

The suppression of creativity within the military can stem from educational models that prioritise conformity and obedience over fostering creative thinking. Although schools play a crucial role in cultivating creativity, the military's emphasis on training and education often reinforces rigid structures and constrains innovative thinking. As individuals progress through the educational system and enter the military, the focus shifts towards conformity, leading to a gradual erosion of creativity (Robinson, 2006; Zweibelson, 2024).

Metacognitive awareness raising

Learning about how we learn, understanding how we understand, and delving into lessons on brain function are critical aspects of fostering innovation. By enhancing cognitive understanding through such practices, individuals can unlock greater creativity and problem-solving abilities. However, within defence contexts, there's often a lack of understanding regarding decision-making processes and their underlying reasons. Without this insight, it becomes challenging to identify and address the barriers to innovation effectively. Therefore, by increasing metacognitive awareness, defence organisations can better comprehend the factors that inhibit innovation, paving the way for more effective strategies to foster creativity and adaptability (Kim & Lee, 2018).

Functional fixedness

Functional fixedness acts as a blocker to innovation in defence by limiting individuals' ability to perceive alternative uses for existing resources or technologies. This cognitive bias prevents individuals from considering innovative solutions to problems because they are unable to see beyond the original purpose of a given tool or resource. In defence contexts, where resourcefulness and adaptability are crucial, functional fixedness stifles creativity and inhibits the exploration of novel approaches to challenges. By constraining the ability to think outside the box and repurpose existing assets, functional fixedness hampers the military's capacity to innovate and respond effectively to evolving threats and circumstances (McCaffrey, 2012).

(Continued)

Table 5.1 (Continued)

___

Military Innovation Constraints

___

Organisational resource availability and scarcity	**Resource allocation** *Budgetary constraints and limited resources can significantly hinder the ability of units to invest in research, development, and experimentation. Without adequate funding, military units may struggle to pursue innovative initiatives or acquire cutting-edge technologies, ultimately impeding their capacity to stay ahead of emerging threats and maintain operational superiority. Similarly, time constraints and competing priorities limit the availability of dedicated resources for innovation within military units. When personnel are overwhelmed by immediate tasks and operational demands, they may prioritise short-term objectives over long-term innovation efforts, resulting in a lack of time and resources allocated to creative endeavours and strategic planning* (Amyx, 2019; Groysberg et al., 2018). **Outdated infrastructure** *In spaces lacking adequate training and tools to foster creativity, personnel may struggle to generate innovative ideas. Legacy systems further compound this challenge, posing compatibility issues with emerging technologies and requiring costly modernisation efforts. With a scarcity of collaborative and innovative spaces, the military may struggle to adapt and leverage new capabilities effectively* (Rabelo & Bernus, 2015). **Funding** *A lack of discretionary funding can significantly impede military innovation by limiting resources available for research, development, and experimentation. Without adequate funding to support innovative projects and initiatives, the military may struggle to invest in emerging technologies, explore new concepts, or incentivise creative thinking among personnel. Additionally, the absence of funding specifically allocated for innovation may discourage individuals from pursuing innovative ideas, as there may be little incentive or support for such endeavours* (Fan et al., 2019; Yigitcanlar et al., 2018). **Organisational support (champions and sponsors)** *Organisational support, including the provision of champions and sponsors, is vital for fostering military innovation. Good champions exhibit proactive leadership and inspire others, while the absence of such support can hinder innovative initiatives from gaining traction and realising their potential* (Read, 2000; Artto et al., 2008; Howell et al., 2005; Kelley & Lee, 2010). **Industry support (availability or lack of)** *The military often relies on civilian industries for cutting-edge technology and innovation. For example, the US heavily relies on partnerships with organisations like NASA for technological advancements. However, industry partners may withhold technology, hindering military innovation by limiting access to crucial resources and capabilities* (Brenk, 2020).

(Continued)

Table 5.1 (Continued)

Military Innovation Constraints

Individual leader cognitive attributes	**Stress, decision fatigue, and cognitive load** *Self- and institutionally generated stress, including cognitive load, decision fatigue, physical fatigue, and glucose consumption, can all act as significant blockers to military innovation. When personnel are overworked or under extreme stress, cognitive function is impaired, leading to reduced creativity, problem-solving ability, and decision-making capacity. Decision fatigue further compounds these challenges as individuals become less able to evaluate and prioritise innovative ideas effectively. Additionally, physical fatigue can degrade physiological functioning, impacting cognitive processes essential for innovation. Glucose consumption and amygdala hijack exacerbate these effects, impairing rational decision-making and hindering the ability to generate and implement innovative solutions to military challenges* (Sweller, 1988; Lindsay et al., 2009; Noworol et al., 2017; Rock, 2009; Dienel, 2019; LaManna et al., 2009; Mergenthaler et al., 2013). **Intelligence and neural activity** *When individuals face challenges in processing and analysing information due to cognitive limitations or neural inefficiencies, it can hinder their ability to generate innovative solutions to complex problems. Factors such as information overload or cognitive biases can further impede the cognitive processes necessary for innovation* (Neubauer & Fink, 2009; Khalil et al., 2019; Miron et al., 2004; Sprugnoli et al., 2017). **Predilection towards reductionism** *The predilection towards reductionism in the military, where complex warfare is often simplified into manageable components, can act as a blocker to innovation efforts. By relying on reductionist approaches in understanding and addressing complex challenges, the military may overlook nuanced factors and fail to grasp the intricacies of modern warfare fully. This can inhibit innovation by limiting the exploration of holistic and interdisciplinary solutions that are better suited to navigating the complexities of contemporary security environments* (Edwards, 2022). **Lack of collaboration** *When individuals or units prioritise competition over collaboration, it can lead to siloed thinking, a reluctance to share information, and a lack of cooperation across organisational boundaries. Similarly, a culture of individualism may discourage teamwork and collective problem-solving, as individuals focus on personal achievements rather than collective goals. This can result in missed opportunities for leveraging diverse perspectives and expertise to generate innovative solutions to complex challenges* (Taylor & Wilson, 2012).

(Continued)

Table 5.1 (Continued)

Military Innovation Constraints

Expert advice

Expert advice, when improperly utilised, can act as a blocker to innovation, particularly when experts dominate conversations and dismiss alternative perspectives. In a design workshop, experts may stifle creativity by insisting on their own solutions, hindering the exploration of diverse ideas and approaches. Instead, experts should be integrated into the panel to provide valuable insights while allowing space for collaboration and the consideration of multiple viewpoints (Chandy & Tellis, 2000).

Minimum viable product

The misconception that a minimum viable product (MVP) must be certain to succeed before testing can hinder innovation efforts. The essence of MVP lies in learning through testing, not in proving or validating preconceived notions. Embracing uncertainty and using testing as a means to learn and iterate is crucial for fostering innovation and ensuring the development of successful products or solutions (Anderson et al., 2017).

Type 1 vs. Type 2 thinking (fast and slow)

The reliance on Type 1 thinking, characterised by quick, instinctive responses, over Type 2 thinking, which involves deliberate, analytical reasoning, can hinder military innovation. Military cultures often prioritise rapid decision-making and action, favouring Type 1 thinking over the more deliberate and reflective Type 2 approach. However, this can lead to the adoption of mental models that are not thoroughly scrutinised or challenged, potentially overlooking nuanced aspects of complex military challenges (Kahneman, 2011; Soni, 2021).

Preference for simplicity and stability

The military's preference for simplicity and stability, often favouring straightforward plans and solutions, can hinder innovation efforts by promoting a risk-averse mindset and limiting the exploration of complex strategies. While simplicity can reduce the likelihood of errors, it may also overlook nuanced aspects of military operations and hinder adaptability in rapidly changing environments. The concept of "soldier proofing," aiming for simplicity to minimise potential mistakes, can further reinforce this tendency (Leonard, 2022; Zweibelson, 2022).

The law of least effort (Glucose Preservation)

Individuals naturally gravitate towards the path of least resistance, but innovation often requires additional effort, experimentation, and risk-taking, which can be perceived as burdensome or unnecessary by those accustomed to taking the easiest route. Consequently, a reluctance to invest time and energy into innovative endeavours may prevail, hindering the adoption of new ideas and the pursuit of creative solutions to military challenges (Dienel, 2019; Rock, 2009; LaManna et al., 2009; Mergenthaler et al., 2013).

(Continued)

Table 5.1 (Continued)

Military Innovation Constraints

Time to think

The lack of dedicated time for deep thinking and reflection can indeed act as a significant blocker to military innovation. Without sufficient downtime for mental rest and rejuvenation, personnel may experience mental fatigue, which reduces cognitive capacity and hampers their ability to innovate effectively. The absence of structured time for reflection and contemplation limits opportunities for generating new ideas and exploring innovative solutions to military challenges (Karjalainen et al., 2006).

Introverts vs. Extroverts vs. Ambiverts

Individuals who lack social courage or confidence may struggle to contribute effectively to innovation efforts. In a culture that values open communication, collaboration, and the challenging of assumptions, those who are hesitant to speak out or engage in social interactions may find it challenging to participate fully in innovation initiatives (Hosseinzadeh & Yoosefi, 2022).

Cognitive and allostatic load

Cognitive load, including allostatic load, encompasses three types of information management challenges that impact military innovation. First, individuals receive an overwhelming amount of information, necessitating efficient processing to avoid cognitive overload. Second, managing the flow of information requires effective allocation of cognitive resources to prioritise tasks and maintain focus amidst distractions. Lastly, stress plays a critical role, as heightened stress levels can impair cognitive function, particularly in multitasking situations (Cheng et al., 2020).

Professional ideology

Professional ideology, market ideology, and bureaucratic ideology represent three distinct perspectives that influence innovation within the military. Professionals adhere to a belief in doing what is right to accomplish the mission, prioritising effectiveness and excellence in their work. Conversely, market ideology emphasises business principles and financial incentives, potentially promoting innovation that aligns with profit motives. Bureaucratic ideology focuses on adherence to rules and regulations, which may stifle innovation by prioritising procedural compliance over creative problem-solving. Fostering a professional ideology is essential for promoting innovation within the military. This ideology prioritises doing what is right to achieve mission success, emphasising effectiveness, excellence, and a commitment to continuous improvement. By cultivating a culture that values professionalism, integrity, and a relentless pursuit of excellence, the military can create an environment where innovation flourishes. Encouraging personnel to embrace professional values and ethics while empowering them to pursue innovative solutions to complex challenges is key to driving meaningful innovation and maintaining strategic advantage on the battlefield (Bentley, 2005; Simons, 2008).

(*Continued*)

Table 5.1 (Continued)

Military Innovation Constraints

Intrinsic vs. Extrinsic motivation

Intrinsic motivation, driven by internal factors such as personal satisfaction or a sense of purpose, often leads to more sustained and meaningful engagement in innovation efforts. However, when innovation is primarily incentivised by extrinsic factors such as rewards or recognition, it may lead individuals to focus on meeting external expectations rather than pursuing innovative solutions that align with strategic objectives. This can result in a lack of genuine passion or commitment to innovation, hindering the organisation's ability to adapt and thrive in a rapidly changing environment (Aalbers et al., 2013; Bhaduri & Kumar, 2011; Fischer et al., 2019).

Dunning–Kruger effect

The Dunning–Kruger effect, where individuals with limited knowledge or expertise overestimate their abilities, can act as a significant blocker to military innovation. When personnel are unaware of their own limitations or overconfident in their capabilities, they may resist seeking input from others or exploring alternative perspectives, thus hindering collaboration and innovation (Vergauwe et al., 2018).

Individual leader character attributes (reinforced)	**Innovation in error-producing conditions**

Innovation in error-producing conditions, such as those found at the edge of chaos, presents unique challenges characterised by ambiguity, instability, and unknown variables. In such environments, traditional approaches may not yield predictable or safe solutions. Instead, innovation necessitates the use of abductive thinking, which involves inferring the best possible explanation based on incomplete information, and design methodologies that emphasise experimentation and iteration. By embracing uncertainty and leveraging creative problem-solving techniques, the military can navigate error-producing conditions and drive innovation in strategy, operations, and technology (Williams, 2015).

Absence of cultural leadership

The absence of a culture that actively fosters and values creativity can significantly hinder operational effectiveness. When commanders neglect to emphasise the importance of innovative thinking, it may result in personnel hesitating to propose novel solutions or challenge conventional methods, thereby jeopardising mission success. Rigid command structures within military hierarchies can impede innovation by stifling initiative and restricting autonomy among junior personnel. The overly centralised decision-making authority often discourages lower-ranking individuals from suggesting alternative approaches or deviating from established protocols. Excessive micromanagement that is prevalent in military units further exacerbates this issue by inhibiting independent thinking and initiative. When leaders closely supervise subordinates and dictate every aspect of their tasks, personnel become hesitant to take risks or

(Continued)

Table 5.1 (Continued)

Military Innovation Constraints

propose innovative solutions, fearing potential reprimand or disapproval. Resistance to unconventional ideas persists within military environments that prioritise conformity and adherence to established procedures. The fear of criticism or ridicule within military units can deter individuals from expressing their creative ideas or proposing alternative solutions. When personnel anticipate negative repercussions for deviating from established practices or challenging authority, they may opt to conform to existing norms rather than risk potential censure (Stewart, 2009).

Arrogance/Ego/Humility

Arrogance, ego, and a lack of humility create barriers to collaboration, hampering open communication and inhibiting the acceptance of new ideas or feedback. When individuals or groups within the military are driven by arrogance or ego, they resist alternative perspectives or innovative approaches, leading to a culture where innovation is marginalised or dismissed. A lack of humility can prevent individuals from acknowledging their own limitations or learning from failures, further impeding the iterative process necessary for innovation to thrive (Zweibelson, 2015).

Personnel dynamics

Limited access to education or professional development opportunities creates a void in requisite knowledge and skills to engage in innovative thinking; consequently, individuals encounter difficulties in devising creative solutions to complex challenges or adapting to evolving operational requirements. The impetus for innovation can also be stifled when personnel grow complacent and overly accustomed to existing practices in military units and/or environments. Resistance to change and a reluctance to explore new approaches hinder organisational adaptability and agility, ultimately hampering the pursuit of innovative solutions (Dapra et al., 1985).

Imposter syndrome

Imposter syndrome describes a phenomenon where individuals doubt their own abilities and feel like they don't belong or deserve their position. Those experiencing imposter syndrome may hesitate to share their ideas or take on leadership roles, fearing that they will be exposed as frauds or inadequately prepared. This can lead to missed opportunities for contributing innovative perspectives and solutions to military challenges (den Besten, 2015).

Faulty risk perception/Risk aversion

When individuals or organisations perceive risks inaccurately or overestimate the potential negative consequences of innovative initiatives, they are hesitant to pursue new ideas or approaches. This leads to a reluctance to embrace change or take calculated risks, inhibiting the exploration of innovative solutions to military challenges. Additionally, excessive risk aversion can hinder creativity and limit the organisation's ability to adapt and respond effectively to evolving threats and opportunities (Odell, 2022).

(Continued)

Table 5.1 (Continued)

Military Innovation Constraints

Fear of failure and lack of confidence

These traits influence an individual's mindset, motivation, and willingness to take risks in pursuing new ideas or approaches. Individuals who possess a strong fear of failure are more hesitant to step outside their comfort zones. Similarly, those with low confidence may doubt their abilities to generate creative solutions or navigate uncertain situations effectively, further inhibiting their innovation potential. The fear of failure looms large as a barrier to creativity within military units. When personnel are apprehensive about the potential consequences of failure, such as facing negative repercussions or scrutiny, they may shy away from taking risks or proposing unconventional solutions (Clarke et al., 2011).

Lack of courage and fearlessness

Individuals who lack courage or fearlessness are unwilling to take risks or challenge the status quo, fearing potential failure or negative consequences. This reluctance to step outside their comfort zones or embrace uncertainty stifles creativity and impedes the exploration of new ideas or approaches. Conversely, individuals who possess courage and fearlessness are more likely to embrace innovation, viewing challenges as opportunities for growth and success (Chaleff, 2010; Koerner, 2014).

Ultracrepidarianism

The habit of giving opinions and advice on matters outside of one's expertise can be a personal attribute that blocks innovation. When individuals assert their opinions without sufficient knowledge or expertise in a particular area, it can hinder constructive dialogue and the generation of innovative ideas. This spreads misinformation, or ill-informed opinions may lead to misguided decision-making and ineffective problem-solving (Villain, 2020).

Procrastination

Delaying the exploration of new tactics, technologies, or strategies can result in missed opportunities to enhance readiness, adaptability, and mission success. Procrastination in decision-making processes can lead to inefficiencies, reduced agility, and compromised situational awareness on the battlefield. In a dynamic and rapidly evolving security environment, military organisations must prioritise proactive and timely innovation to maintain a competitive edge and address emerging threats effectively (Haesevoets et al., 2022).

Normalisation of deviance

The normalisation of deviance, an individual characteristic, can serve as a significant blocker to innovation, particularly in safety-critical environments such as the military. This phenomenon occurs when individuals or groups gradually accept deviant behaviours or practices as normal, despite being inconsistent with established standards or protocols. In the context of innovation, the normalisation of deviance leads individuals to

(Continued)

Table 5.1 (Continued)

Military Innovation Constraints

---

overlook safety risks or cut corners in the pursuit of efficiency
or expediency. This reluctance to adhere to established safety
protocols or best practices can compromise safety, increase
the likelihood of errors or accidents, and inhibit the adoption
of innovative solutions that prioritise safety and reliability
(Kern, 2009).

Low signal-to-noise ratio

When individuals are bombarded with excessive information
or distractions, it becomes challenging to discern valuable
insights or identify meaningful patterns amidst the noise. This
plethora of information leads to cognitive overload, reduced
clarity of thought, and difficulty in seeing the big picture or
identifying innovative solutions. Overcoming a low signal-to-
noise ratio requires individuals to develop strategies for filter-
ing information, prioritising relevant data, and maintaining
focus on key objectives (Kern, 2011).

Reinforcing reductionism

When individuals rely solely on past experiences or traditional
mental models to make future decisions, they limit their ability
to explore new possibilities or adopt innovative approaches.
This tendency to anchor decision-making in familiar patterns
or incremental changes, despite knowing their limitations or
past failures, perpetuates reductionist thinking and inhibits the
pursuit of transformative solutions. By prioritising familiarity
over novelty, organisations may miss out on opportunities for
breakthrough innovation and fail to adapt to evolving challenges or opportunities (Rosenberg, 2001).

Poor communication

When communication channels are unclear, inconsistent, or ineffective, it can hinder the sharing of ideas, collaboration among
team members, and the alignment of efforts towards innovation
goals. Misunderstandings, misinformation, and a lack of clarity can lead to confusion and conflict, impeding progress and
stifling creativity. Additionally, poor communication can result
in missed opportunities to leverage diverse perspectives, expertise, and insights, limiting the potential for innovative solutions
to emerge (Larsen, 2011).

Design gaze

The design gaze can act as a blocker to innovation by superficially employing design techniques without a deep understanding of or internalisation of their essence. This phenomenon
involves individuals or organisations that adopt design thinking
methods without genuinely embracing the mindset attributes
of effective designer thinkers. Instead, they may resort to mere
verbal descriptions or token gestures of innovation, such as

(Continued)

Table 5.1 (Continued)

Military Innovation Constraints

hosting short workshops or inviting external consultants, without truly engaging in meaningful divergent thinking or incorporating stakeholder input. This superficial approach not only fails to drive genuine innovation but also perpetuates conservative mindsets and consultancy syndrome, where the focus is on appearance rather than substantive change (Heltberg, 2022; Mehrotra, 2022).

One-way decision gates

Where decisions are made unilaterally and are not subject to revision can be a significant blocker to innovation. This approach curbs creativity and inhibits the ability to adapt to new information or changing circumstances. When decision-makers are unwilling to reconsider their choices, even in the face of strong evidence or consequences, it restricts the exploration of alternative solutions and limits the potential for innovative ideas to emerge. This rigid decision-making process can lead to missed opportunities, increased resistance to change, and a lack of agility in responding to dynamic challenges (Kern, 2009).

Fear of uncertainty/unknown (links to metrics)

The fear of uncertainty and the unknown often leads individuals to cling to numbers and spreadsheets, seeking rigid planning and calculations to mitigate perceived risks. However, this emphasis on quantifiable data can sometimes lead to the illusion of accuracy, with individuals manipulating numbers to fit predetermined outcomes. Overcoming this reliance on certainty metrics requires embracing uncertainty as inherent to complex systems and fostering a culture that values adaptability, resilience, and informed decision-making in the face of ambiguity (Jalonen, 2012; Shane, 1995; Wang et al., 2011).

Luddites

Historically associated with resistance to technological advancements, Luddites often fear that innovation and automation will lead to job loss or render their skills obsolete. Their opposition to new ideas and technology stems from concerns about personal relevance and economic stability (Mellor et al., 2015).

Combat masculine warrior culture: Macho

Rooted in a historical emphasis on aggression and dominance, this culture often prioritises kinetic operations and direct combat over alternative approaches like diplomacy or peacekeeping. Consequently, innovative strategies that diverge from this traditional warrior ethos may face resistance and dismissal as being "woke" or irrelevant. Overcoming this barrier necessitates challenging entrenched attitudes and promoting a broader understanding of military effectiveness that includes innovative solutions beyond conventional warfare (Dunivin, 1994).

(Continued)

Table 5.1 (Continued)

Military Innovation Constraints

Machiavellianism and deference towards kinetic effects

Machiavellianism, characterised by a focus on power and manipulation to achieve desired outcomes, often manifests as deference towards kinetic effects in military contexts. This perspective prioritises aggressive or forceful actions, such as direct combat or kinetic operations, as the primary means of exerting influence or achieving objectives. Consequently, when faced with innovation efforts that propose alternative approaches or emphasise non-kinetic solutions, individuals influenced by Machiavellianism may be dismissive or resistant, viewing them as less effective or insufficiently assertive. Overcoming this bias requires promoting a broader understanding of military effectiveness that encompasses a range of capabilities, including diplomacy, information operations, and humanitarian assistance, alongside kinetic effects (Wadham & Connor, 2023; Whitt & Perazzo, 2018).

5.4 Military Organisational Conditions for Innovation

Western militaries confront novel challenges in contemporary warfare contexts, necessitating continual and critical discourse on the nature and configuration of their organisational structures, conditions, and operational modalities for future endeavours. While advocates of military design thinking accentuate the principles and methodologies intrinsic to their approach, scant attention has been paid to the requisite organisational conditions within military command units for fostering innovation with lasting impact.

As delineated in Chapter 3, extant literature underscores pivotal conditions indispensable for the effective implementation of military design thinking. These fundamental organisational prerequisites stem from exhaustive research conducted by Wrigley et al. (2020). Their comprehensive study, spanning diverse sectors and industries, meticulously dissected the sustainable attributes of design interventions essential for nurturing innovation. These conditions, tailored specifically for the military setting, encapsulate the following considerations:

- **Long-term military strategic vision:** The overarching plans and strategic goals set by military leadership to achieve desired outcomes over an extended period of time, typically involving considerations of threats, resources, alliances, and geopolitical factors.
- **Base infrastructure:** Encompasses not only the physical spaces where creative, design, and innovation activities occur but also the resources and support structures necessary for military operations and readiness to be built.
- **Creative cognitive capability:** The collective expertise, awareness, and proficiency of military personnel in innovation-related activities and initiatives within the organisation.

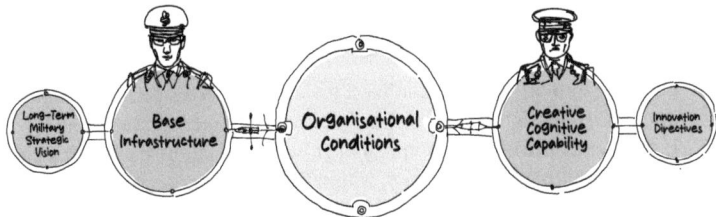

Figure 5.1 The Four Conditions

- **Innovation directives:** The formal instructions or orders issued by military leadership to encourage and ensure the utilisation of innovation, design, and creativity within the organisation's operations. These directives typically outline specific expectations, goals, and accountability measures for integrating innovative approaches into various aspects of military activities and processes.

These four conditions (Figure 5.1) are introduced in greater depth in the following sections. In the following sections, leaders are presented with reflective questions that encourage them to consider how to effectively establish these four organisational conditions within their own units of command.

5.4.1 Long-Term Military Strategic Vision

This condition, encapsulating the long-term strategic intent of an organisation, is crucially tied to the establishment of its strategic direction and forward-thinking approach. Most Western militaries, tasked with defending national interests and contributing to global security, must strike a balance between present operational needs and future requirements. It is imperative for these organisations not to be consumed solely by immediate concerns but also to allocate resources and attention to future horizons. However, crafting a strategic vision for the long term is often challenging, as it requires foresight and a proactive mindset. An illustrative example of this forward-looking perspective can be seen in the actions of historical figures like Alexander the Great, who, recognising the need to ensure the continuity of military strength, strategically won the hearts and minds of defeated nations in such a way that they joined his cause. This meant he had an enduring supply system for both reinforcements and logistics. Such a proactive approach underscores the importance of setting up systems and initiatives that extend beyond the organisation's current lifetime, ensuring its continual relevance and effectiveness in safeguarding national interests and global security.

The presence of such a long-term vision can be established through the following questions:

- Does the military organisation have a clearly defined strategic vision that outlines its mission, objectives, and desired outcomes for the future?
- Is there evidence of a commitment to growth, adaptation to emerging challenges, and fostering innovation within the military organisation?
- How effectively is the strategic vision communicated and understood across all levels of the organisation, from top leadership to frontline personnel?
- Are there mechanisms in place to align day-to-day activities and decision-making with the overarching goals and priorities set forth in the strategic vision?
- Is there an ongoing evaluation and refinement of the strategic vision in response to changes in the geopolitical landscape, technological advancements, and evolving threats?
- How well does the strategic vision integrate with broader national security objectives and align with the priorities set by higher military and governmental authorities?
- Are there measures in place to ensure accountability and track progress towards achieving the strategic objectives outlined in the vision as it evolves and changes over time with changes in leadership?
- What mechanisms exist to ensure that non-competing priorities are included in the different services' vision and command?

5.4.2 Base Infrastructure

The condition of facilities refers to the physical spaces and resources dedicated to design and innovation initiatives by the organisation. The notion of a physical environment required to support an emerging endeavour (e.g., an organisational design hub) being paramount for the success of such an initiative is documented in the literature. The establishment of facilities like CyberWorx (Figure 5.2) exemplifies the importance of providing a conducive environment for such endeavours, with features such as human-centred design approaches, public partnerships, prototyping capabilities, and advanced technology. CyberWorx, operated by the US Air Force, serves as both a facility

Figure 5.2 The CyberWorx's Approach to Human-Centred Design (Preston et al., 2019)

and a training institution, boasting state-of-the-art amenities, immersive labs, and a focus on various domains, including robotics, cybersecurity, data visualisation, and cross-domain integration.

The innovation environment can be established through the following questions:

- Are there dedicated spaces within military bases for innovation activities?
- Do these spaces facilitate collaboration and creative thinking among personnel?
- How accessible are these spaces to personnel across different ranks and departments?
- Does the organisation allocate sufficient funding for design and innovation initiatives?
- Are there designated budgets for acquiring necessary tools and technologies?
- Are personnel adequately trained in utilising resources for innovation?
- Are there established innovation hubs or centres within military bases?
- How well-equipped are these hubs for supporting innovation activities?
- Is there a culture of collaboration and cross-functional interaction within these spaces?
- Are measures being taken to modernise infrastructure and upgrade equipment for innovation needs?
- Is there a designated design facilitator to support the utilisation of these spaces, ensuring they are actively used for addressing complex systems and their innovative outcomes?

5.4.3   Creative Cognitive Capability

This relates to the organisation's people, specifically whether they are all homogeneous thinkers or if they understand the value of innovation and are capable of practising it. Innovation is rarely an individual task. Leonard and Sensiper explain the notion of "knowledge walking out the door" as tacit knowledge leaving alongside the employees who harbour it (1998, p. 112). Such knowledge is essential to the innovation process, yet maintenance of this knowledge is relatively unexplored and absent in the military. Indeed, the success of a unit or command is contingent on its people. For a unit to be considered competent in any given skill, its people are required to be capable of actioning that skill and understanding the value that the skill brings to the organisation (Nusem et al., 2017). Innovation occurs when the impossible meets necessity. Countries facing an existential (or perceived existential) threat have their backs against the wall where failure is not an option. Recognising that conventional methods likely contributed to their current situation, they are generally more open to exploring new approaches. Thus, their priorities change from being a conservative and legacy-driven system prone to institutional inertia to one that has a greater appetite for risk when it comes to exploring more options.

Figure 5.3 Cognitive Creativity Wall

Countries that have long felt vulnerable are those that are often recognised globally as open to greater innovation. While opinions on the list might vary, some examples include small states such as Singapore, Taiwan, and Israel (Raska, 2015). Others might include Ukraine, Estonia, and South Korea. The problem for many other nations that have enjoyed the luxury of more stable regions is that technology and global interdependency mean nonlinear tipping points can create Black Swan events that leave them vulnerable. In the same way that force structures take time to develop, so too, do mindset shifts. For many countries, such responsiveness to highly dynamic changes in the operating environment has not occurred since World War II. Looming threats of great power conflicts mean that complacency in promoting innovative mindsets could be sealing their fate. Collaborative engagements, as shown in Figure 5.3, help tackle this issue by fostering an environment where ideas can be openly shared and displayed. The wall covered with design activities visually represents a collective brainstorming process, where team members contribute perspectives and solutions. This setup enables participants to view and build on each other's ideas, which helps reduce cognitive blind spots by encouraging diverse viewpoints. By showcasing ideas in this way, the organisation can gauge its creative cognitive capability and harness the full spectrum of team insights.

Creative cognitive capability in an organisation can be gauged through the following questions:

• Are all military personnel encouraged and empowered to contribute to innovation initiatives within the organisation?

- Is there a culture that values and prioritises innovation and creative thinking across all ranks and departments?
- How effectively does the organisation capture and retain tacit knowledge essential to the innovation process, especially as personnel transition in and out of roles associated with short-term posting cycles?
- Is there a recognition of the collective expertise and awareness of military personnel in design-related activities, and are efforts made to cultivate and leverage this knowledge?
- To what extent does the organisation's rank-driven mentality and hierarchical culture impede open communication and collaboration, particularly regarding innovation?
- Are there mechanisms in place to promote cross-rank collaboration and ensure that junior ranks have opportunities to contribute their ideas and perspectives?
- How entrenched is the culture of legacy within the organisation, and to what extent does it hinder the adoption of new technologies and approaches?
- Are there initiatives in place to overcome resistance to change and foster a culture that embraces experimentation?
- How does the organisation address the fear of failure, rejection, and loss of control among personnel and encourage them to take risks and propose innovative ideas?
- What measures are taken to mitigate the risk-averse nature of the military environment and create an atmosphere conducive to experimentation and creativity?

5.4.4 Innovation Directives

Having directives denotes that the organisation's people (not just a select few) are mandated to practise innovation – whether through personal performance metrics or through legislation in government – and are held accountable for addressing this mandate. In entire services, the only way to change general practice is by changing the processes by which it is done. The process of managing new ideas into new practices so that innovative solutions are implemented and institutionalised comes largely down to the directives set. In the absence of a robust system that recognises and rewards innovation, personnel may hesitate to invest time and effort into pursuing innovative projects. However, just as the following example explains, setting directives and mandates is a fine balance. Frederick William I, the king of Prussia (Figure 5.4), was not known for having a pleasant disposition. His passion was his army, and he spent much of his life building it. Frederick often walked the streets of Berlin alone, and his subjects fled from him. It is said that on one of his walks, a citizen saw him coming and attempted to escape the monarch by ducking into a doorway.

"You!" called out the king. "Where are you going?"
"Into the house, Your Majesty," replied the nervous man.
"Is this your house?" Frederick pressed.
"No, Your Majesty."
"Then why are you entering it?" the king demanded.
"Well, Your Majesty," the man admitted, worried that he might be
thought a burglar, "to avoid you."
"Why?" demanded Frederick.
"Because I fear you, Your Majesty."
Frederick raised his walking stick threateningly at the man and shouted,
"You're not supposed to fear me. You're supposed to love me!"

The delicate balance between setting directives and nurturing innovation
underscores a pivotal aspect of organisational management. Excessive imposition of governing structures may inadvertently hinder the emergence of
novel ideas and impede the proactive engagement of staff members tasked
with catalysing innovative initiatives. This necessitates adept leadership with
a nuanced approach, wherein the imposition of directives is tempered by the
cultivation of an environment conducive to creativity and entrepreneurial

Figure 5.4 Frederick Directives

behaviour. Effective leadership (the opposite of King Frederick) in this context entails the establishment of a culture that not only tolerates but also actively encourages experimentation, embraces risk-taking as a fundamental component of progress, and recognises the inherent value of diverse perspectives in fostering innovation.

The presence of directives can be determined through the following questions:

- Are personnel mandated to practise innovation and design as part of their performance metrics or through government directives?
- Is there accountability mechanisms in place to ensure that individuals are held responsible for addressing this mandate?
- How effectively are new ideas transformed into new practices within the organisation, and to what extent are innovative solutions institutionalised?
- Is there a robust system in place that recognises and rewards innovation, or does a lack thereof discourage personnel from investing time and effort into pursuing innovative projects?
- Does a risk-averse culture within the organisation hinder individuals from embracing the necessary risks associated with innovation?
- Is there trust between senior officers and junior counterparts, or does a lack of trust stifle innovation within the organisation?
- Is there a supportive environment that nurtures open dialogue and collaboration, or does a hierarchical structure hinder creative thinking and innovation?
- Are individuals empowered to share their ideas and take initiative, or do traditional power structures limit the potential for diverse perspectives and innovative ideas to flourish?
- Is there a clear mandate or organisational support for innovation, or do individuals struggle to prioritise creative thinking without explicit encouragement?
- Are innovative efforts recognised and rewarded within the organisation, or does a lack of motivation inhibit engagement in creative endeavours?

In closing, military organisations must navigate a unique blend of structural and psychological constraints that challenge their capacity for innovation. Four key organisational conditions are essential for cultivating a more adaptive, innovative ecosystem: fostering dedicated collaborative spaces, ensuring accessibility across ranks, committing resources, and promoting a culture of cross-functional interaction. Leaders are encouraged to reflect on these conditions through specific questions, assessing how these elements can be effectively implemented within their units. Embracing these strategies better prepares military organisations for the evolving complexities of modern warfare, building a culture that actively fosters calculated risk-taking and responsive innovation.

6 Towards New Horizons

Don't walk backwards into the future.

6.1 Embracing Complexity

Turning to the future, this final chapter draws together the key themes outlined so far and offers general concepts for those in leadership roles. As with any complex adaptive system, though, there is no one silver bullet to fix everything. The suggestions offered here are general principles to guide decision-makers in adapting options to their organisation's unique setting. Nurturing a culture of innovation is not a problem to be solved; rather, it is a living, breathing organism that requires constant adjustments to refocus gently on a more desirable state.

In contrast to the structured approach suited for complicated problems that militaries are adept at addressing, complexity demands a distinct methodology. Rather than relying on preconceived strategies, navigating complexity necessitates an active engagement with the system to discern its responses. Through this iterative process, crucial relationships and dependencies emerge, guiding subsequent actions.

Despite the temptation of adhering to conventional military practices ingrained through historical precedent, the proliferation of complex adaptive systems mandates a paradigm shift. Embracing creativity becomes imperative in navigating the complexities of the contemporary world. Ultimately, the future will favour those most adaptable to change.

6.2 Redefining Success

Success is not defined by a single event; rather, it is thriving in a constantly dynamic system. While tactical-level achievements, such as securing a beachhead, might give the illusion of success, these are just enablers to not only win a war but also set the conditions for enduring peace long after the guns fall silent. Thus, although militaries will continue to protect their junior personnel

DOI: 10.4324/9781003502180-6
This chapter has been made available under a CC-BY-NC-ND 4.0 license.

with the illusion that complicated processes solve problems, the reality is that senior decision-makers need to mature past the security blanket mindsets that single-shot inputs yield enduring solutions. To reinforce an earlier point – complex systems are constantly evolving and cannot be solved; rather, they are only influenced in more favourable states, requiring ongoing interventions.

This does not mean issuing orders to those on a two-way range should include complexity theory, but it does mean any thinking used to derive the plan must resist oversimplifying the system, hence the maxim, "Simplicity at the edges, complexity at the core." The opposite is also true when planning treats the system as only simple – chaos ensues. Hence the refrain, "Complexity at the edges, if simple at the core."

Being comfortable with complexity is challenging. It demands enhanced cognitive agility acquired through innovation mindsets – not only in the decision-maker but also from their entire team of advisers and planners. These people do not magically become "anticipative geniuses" thriving in complex systems simply because they are posted into a new assignment. Just like the physical fitness to run a marathon requires many hours of training, developing innovative mindsets takes many years to nurture. This is doubly true when the organisation has invested heavily in force-feeding complicated thinking mindsets during their formative years. Recognising military success involves knowing that your organisation is relentless in its pursuit of alternative perspectives success, therefore, is not measured by defeating an enemy on the battlefield; instead, it is every day the military outmanoeuvres its adversaries before a single bullet is fired.

The supreme art of war is to subdue the enemy without fighting.
        – Sun Tzu, *The Art of War*

Developing a deeper culture of innovation means more than just sweeping organisational initiatives; importantly, it also requires investing in individuals. As outlined in Chapters 4 and 5, both top-down and bottom-up initiatives are instrumental in this enduring challenge of maintaining a transient advantage over the adversary. Decades of fighting wars of choice against insurgents have fostered a level of complacency that is ill-suited to developing the innovative mindsets needed to counter a numerically superior peer adversary. As the war clouds grow darker on the horizon, the West must invest in more than just platforms. Being outsmarted in the planning rooms will negate any efforts on the front lines. To maintain the harmony of a rules-based order, it is imperative the liberal-democratic world does more to out-innovate those harbouring expansionist ambitions.

The art of fostering a more innovative culture transcends pre-engineered templates of success. What works in one organisation won't necessarily work in another. Leaders must model the practices they want their staff to adopt. Determining the essence of the bespoke initiatives requires creative thinking, thus practising what they preach. Fortunately, there are some known principles that appear to have greater success than others. They do need adapting and constant re-evaluation as their efficacy diminishes over time, and the wicked system continues to morph and contort. Later sections of this chapter highlight some key principles of Russel Ackoff's dissolution concept of influencing complex adaptive systems (Ackoff et al., 2006). These principles permeate the various models and processes captured under the banner of military design thinking (Chapter 3).

In the journey to redefine success, a profound cultural shift is indispensable. Modern militaries must transition towards a culture that embraces change, creativity, and disruptive mindsets. Promoting an ongoing culture of alternative perspectives should not be narrowly defined; instead, it must encompass adaptability and innovation, fostering an environment where mavericks are celebrated, ideas are explored, and challenges are met with confidence. As shown in Figure 6.1, there are many cultural indicators to assess the transition from that of legacy to one focused on the future. Those who linger on the left-hand side are those who do not truly embrace innovation. As the paraphrased saying goes, "If you don't like change, then you are going to hate irrelevancy."

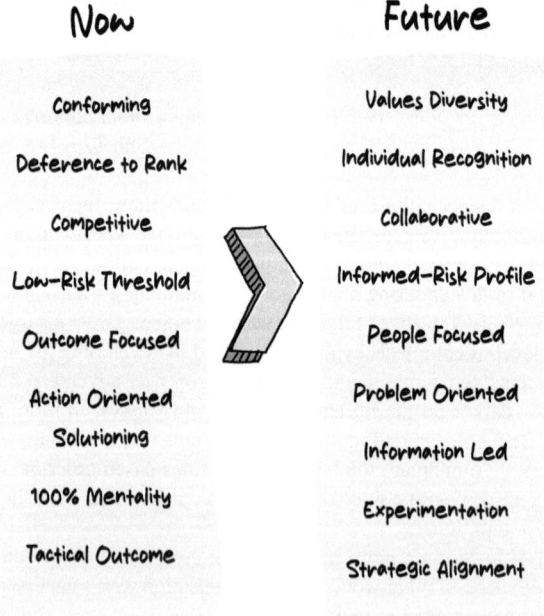

Figure 6.1 Shifting Military Mindsets

6.3   Three Horizons Ahead

The *three horizons framework* is one of many useful tools for implementing cultural change. As presented by Sharpe et al. (2016), it integrates strategic foresight with organisational transformation. By encouraging individuals to articulate their underlying assumptions and contemplate emerging changes, the framework facilitates a reframing of perspectives regarding objectives, desires, and actions. In the military context, leveraging this framework enables readiness for future complexities and emerging threats by promoting a transformative approach to challenges, thereby augmenting capabilities and enhancing operational effectiveness.

Effective military preparedness entails not only addressing immediate symptoms of crises but also delving into the root causes embedded within structural and systemic frameworks. Embracing complexity necessitates an acceptance of uncertainty, adaptability to change, and agility in the face of unpredictability. Initiating a profound cultural dialogue within defence institutions fosters a continuous process of inquiry, encouraging the exploration of pertinent questions and the formulation of provisional solutions that drive cultural evolution and ongoing learning. Like other strategic planning tools, the three horizons framework can help militaries cultivate a responsive and resilient posture, poised to navigate the complexities of an ever-evolving security landscape.

This framework offers a structured approach to comprehending the intricate web of interconnected challenges and opportunities that unfold over time (Figure 6.2). The transition from the established norms of the framework's first horizon to the emergence of fundamentally new paradigms in the third horizon is facilitated by the intermediary phase of the second horizon. Within

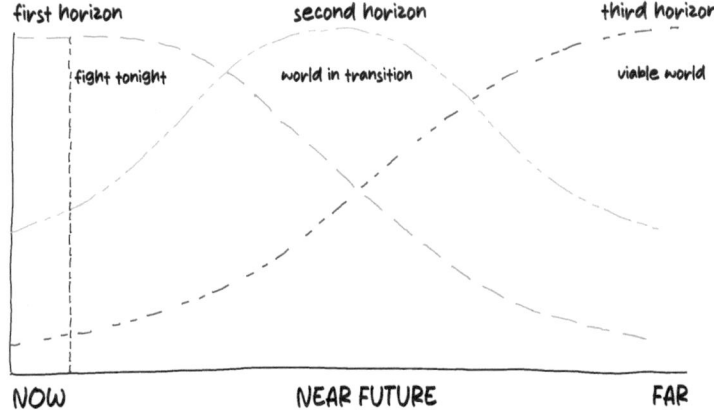

Figure 6.2 The Three Horizons Framework

this transition, certain manifestations of "disruptive innovation" may merely integrate into the existing patterns of the first horizon without inducing significant transformative change. Conversely, other manifestations of disruptive innovation serve as potential conduits, bridging the gap between the established norms of the first horizon and the envisioned future of the third horizon. Although the actual timeframes offered in the three horizons framework might differ from individual country strategic planning (e.g., Australia's National Defence Strategy), the framework's value is more in the concepts than in specific years.

In a military context, Horizon 1 (H1) depicts the current or the ongoing operational environment, often characterised by routine activities and familiar procedures – essentially, "fight tonight." It is where sustaining innovations are implemented to maintain current military operations. At the risk of oversimplifying, many people think of H1 as being what can be achieved within the resources available in the current financial year.

Horizon 3 (H3), on the other hand, represents the aspirational vision of a future operational environment, often depicted as the "viable world" being sought. While this may not possess a detailed blueprint of this future due to its inherent uncertainty, it can help anticipate fundamental transformations and observe emerging social, ecological, economic, cultural, and technological trends. Using the simplified metaphor, H3 is often likened to more future-focused conceptual theories that are ahead of their time but help shape Horizon 2 (H2) strategic thinking today.

H2 serves as the bridge, portraying the "world in transition." Here, innovative ideas and technologies that are already feasible, culturally acceptable, and capable of disrupting and reshaping the current state (H1). These innovations, when embraced, have the potential to catalyse significant changes within military operations, either fostering regeneration, maintaining neutrality, or causing degenerative effects on socio-ecological dynamics. The following summarises the applicability of the framework in defence:

- H3 embodies military cultures capable of constant adaptation and transformation in response to evolving strategic environments and operational challenges.
- Military innovation ecosystems are not static, with no fixed endpoint in achieving and maintaining an H3 scenario indefinitely. Embracing uncertainty and maintaining a learning mindset are crucial during this journey.
- H3 thinking in the military is informed by a new cultural narrative emphasising interdependence and acknowledges the contributions of H1 and H2 perspectives in fostering strategic agility and innovation.
- The three horizons approach values the contributions of each perspective and offers a structured methodology for fostering strategic foresight and embracing adaptability in shaping the future of military operations.

6.4 Leading the Change

As we conclude our exploration of *Creativity in Military Complexity*, the imperative for immediate and decisive action stands clear. The ever-evolving landscape of modern warfare, marked by its unpredictability and intricate challenges, necessitates a progressive shift in our approach to military strategy and operations. The time is now to wholeheartedly embrace innovation, creativity, and divergent thinking within our military institutions. This is not just a matter of keeping pace with technological advancements or potential adversaries; it is about fundamentally rethinking the way we approach complex adaptive systems, make decisions, and plan for the uncertain future.

By fostering participatory epistemic diversity, militaries can break down the silos of conventional thinking, enabling a richer, more varied pool of ideas and perspectives. This diversity is our strongest weapon in combating the inherent cognitive biases that often cloud judgement and hinder effective decision-making.

Our journey through this book has underscored the critical need to integrate nonlinear systems thinking into the very fabric of military planning and execution. Traditional linear models are ill-suited to the complex dynamic environments that characterise contemporary conflicts. So, although the complexity of a system was typically envisaged by the commander, and the complicated planning checks were completed by staff, today, the rate of technological advancements means that the planning-level staff need to be more involved in the complexities of the situation. This has resulted in the legacy transactional processes performed by staff now becoming automated by technology – including artificial intelligence. Their thinking, therefore, needs to elevate above mere complicated checklists. Such a shift requires not only structural and procedural changes but also a cultural transformation within the military. Leaders at all levels must champion this new way of thinking, encouraging experimentation and learning from both successes and failures. Only through such a profound transformation can the West hope to navigate the complexities of modern warfare more effectively and faster than our adversaries. The call to action is clear: it is time to boldly step into a future where creativity and complexity are not just acknowledged but also integral to every aspect of military design and strategy.

6.5 Recommendations for Leaders

In light of the evolving landscape of potential threats, this book is, first and foremost, a conversation starter for those in positions of influence to consider their organisation's unique place on the continuum and to plot their own journey forward. For those who feel that greater fidelity is required, there has been a tentative exploration of what successful organisations are trialling. The principles-based list of potential initiatives centre around empowering military leaders to foster a culture conducive to creativity, design, and innovation within their respective units or spanning the entire military workforce. It is imperative to acknowledge that no singular measure serves as a panacea. Rather, a concerted and multi-pronged approach, collectively embraced by military leaders at every echelon, is poised to yield the most profound impact on our prospective readiness to confront future adversarial engagements.

6.5.1   Leave Your Ego at the Door

The military has a long history of promoting leaders who display confidence and decisiveness. This can lead to a predominance of "Dunning–Krugers" in the organisation, who are more concerned with protecting their image than admitting they are not always right (Dunning, 2011; Vergauwe et al., 2018). Great leaders who are genuine in advocating innovation will be open to ideas from their subordinates and give them a voice. Cyrus the Great's famous approach, "Diversity in counsel, unity in command," might today be referred to as "Disagree in private, but leave with one voice." This maxim refers to the value of holding closed-door sessions where principal advisers offer free and fearless advice, but with the essential caveat that once the decision is made, all remain loyal. Inside the privacy of the counsel, a good principle is "Everyone goes once, seniors go last." This concept means no one is allowed to speak a second time until everyone in the room has contributed, and those who are recognised experts on the matter should always go last. This helps prevent the Dunning–Krugers dominating the discussion and stifling innovation.

At an organisational-cultural level, questions should be asked about what incentives are in place to manage the balance of Dunning–Krugerism. One such thought experiment pondered over by the designers of professional military education courses is the moot question, "Why does the military promote confidence over competence in its officer's corps, yet competence over confidence in its enlisted ranks?" Both the affirmative and negative teams, of course, have plenty of material to play with, but the question is worth reflection. Furthermore, if true, should it? What are the multi-order effects on not only the organisation's culture but also its mission success? Leaders should look to the following for action in the workplace:

- **Foster open dialogue:** Encourage safe expression of ideas by implementing practices like "everyone goes once, seniors go last" in meetings.
- **Adjust incentive structures:** Regularly assess and adjust incentives to balance confidence and competence, reflecting on their impact on culture and mission success.
- **Promote unity in decision-making:** Embrace "diversity in counsel, unity in command" by encouraging honest feedback in closed sessions and ensuring public unity after decisions are made.

6.5.2 Encourage Collaborative Thinking

A collective pool of knowledge and perspectives invariably surpasses the insights of any single individual. Linked to the problem of Dunning–Kruger mindsets is the imperative to cancel cognitive blind spots by maximising the diversity of thought in influencing complex adaptive systems (Dunning, 2011). Decisive individual decision-making is not only vital at tactical levels but also generally effective for complicated problems where established protocols exist. As these leaders climb the career ladder, however, they need to increasingly leverage the collective wisdom of their advisers. Complex systems, by definition, are too overwhelming for one person to comprehend and anticipate – no matter how arrogant they are.

One of the major challenges in defence forces is the pervasive siloed thinking, where individuals rarely look beyond their own discipline, unit, service, or the broader defence system. This isolation extends to the defence industry and even includes active information withholding from those with the necessary clearances. It is crucial to engage the right people and ensure a truly diverse user set, especially the main users for whom solutions are being designed. However, the prevailing attitude is that defence issues are unique and can only be addressed internally, leading to a reluctance to open up and collaborate. This mindset hinders the potential for innovative and effective problem-solving. The following two key suggestions are worth pondering:

- **Leverage collective wisdom:** Encourage collaborative thinking by involving a diverse set of advisers to tackle complex adaptive systems, thereby reducing cognitive blind spots and enhancing decision-making.
- **Break down silos:** Actively promote cross-discipline, cross-unit, and cross-service collaboration within the defence forces and with external partners to foster innovation and effective problem-solving.

6.5.3 Embrace Coopetition vs. Competition

Military organisations are inherently hierarchical, with rank visibly displayed on uniforms and dictating how individuals address each other verbally and behaviourally. This hierarchy drives competition to perform and climb the ranks, sometimes fostering behaviour that undermines the tribal unity necessary for unit cohesion and performance. Members are expected to have each other's backs and adhere to a tribal code of conduct, especially during times of conflict. However, the internal race for advancement can lead to behaviours reminiscent of a sports team with different factions. Instead of focusing on defeating the opposing team, these factions become distracted by internal competition, scheming, and power jockeying, ultimately undermining unity.

Instead of fostering purely competitive environments, there is a need to shift towards "coopetition." This term blends "cooperation" and "competition" to describe a scenario where entities collaborate for mutual benefit while maintaining their competitive edge. This approach is particularly vital when dividing resources such as personnel, assets, and finances, which are often tugged across various services and groups. Without a clear priority path, entities might perceive themselves as more important than their counterparts, leading to delusions of grandeur within the military organisation. To address this, it is recommended that the chief and/or commander provide a lucid priority for our future integrated force and allocate resources accordingly. The following key suggestions are worth considering:

- **Promote coopetition:** Foster an environment where cooperation and competition coexist, encouraging units to work together for mutual benefit while still striving for excellence.
- **Clarify resource allocation:** Ensure a clear, strategic priority for resource distribution across services and units to prevent internal competition from undermining overall military effectiveness.

6.5.4   Reduce the Bureaucratic Shackles

Bureaucratic barriers often stem from a multitude of factors, including historical precedent, risk aversion, and the inherent complexity of military operations. Historical precedent and institutional inertia contribute to the perpetuation of bureaucratic processes that may have outlived their utility but remain ingrained within organisational structures. Moreover, risk aversion, particularly in environments where failure is not tolerated, incentivises the implementation of rigid bureaucratic protocols as a means of minimising potential liabilities. A good indicator of an organisation's risk aversion is the degree to which they legislate and punish well-intended failure.

To mitigate these challenges, military leaders must demonstrate courageous leadership and decisiveness in challenging the status quo by streamlining bureaucratic processes. This necessitates strong leadership and a willingness to accept calculated risks in pursuit of operational agility and innovation. By having the courage to remove unnecessary bureaucratic layers, leaders can streamline decision-making processes and facilitate the rapid approval and testing of new concepts.

Fail early, fail often, but always fail forward.

Key strategies for mitigating bureaucratic hurdles include streamlining the process for approving and testing new concepts, appointing dedicated liaison officers to assist with navigating bureaucratic processes, and regularly reviewing and revising policies that inadvertently stifle innovation. These measures aim to enhance agility and responsiveness within military organisations, thus empowering personnel to explore and implement new ideas with greater efficiency and effectiveness.

However, it is essential to recognise that not all areas are ripe for innovation, and certain domains, such as nuclear power, require stringent bureaucratic oversight to ensure safety and security. Thus, any efforts to streamline bureaucratic processes must be balanced with the need to maintain appropriate levels of oversight and risk management in critical areas. Two key suggestions here include:

- **Streamline bureaucratic processes:** Demonstrate courage by challenging the status quo, removing unnecessary layers, and simplifying approval and testing procedures to enhance operational agility and innovation.
- **Balance oversight and flexibility:** Appoint liaison officers to navigate bureaucratic processes and regularly review policies, ensuring streamlined processes while maintaining necessary oversight in critical areas.

6.5.5 Empower Decision-Making Down

Innovation is widely celebrated on operations but often lamented in peacetime. During the latter, progressive layers of bureaucracy are increasingly added following successive inquiries or unhelpful political interference which is motivated by non-military outcomes. Such changes are often made in isolation for a specific tangible outcome but without sufficient consideration of the wider impact it might have on the military's *raison d'être*. Furthermore, new layers are frequently added, but they rarely fully consider what impacted policies should subsequently be removed or amended. In times of crisis, however, many of the irrelevant handbrakes are removed by empowered commanders when the military is trusted to run the operational level as they see fit. Oversight of implementing national strategy is appropriately retained, but the middle to lower levels of the ends-ways-means trinity are suddenly empowered to make much greater decisions on their own. The challenge of this situation is that the military grows their thinkers with peacetime disempowered mindsets. If, and when, major conflicts break out, decision-makers' performance as innovative thinkers is constrained by many years of Pavlovian responses to disempowered linear thinking. The adage "train as you mean to fight" is as much cognitive as everything else.

The prevalence of bureaucratic hurdles and paperwork within military organisations worldwide presents a pervasive challenge that not only hampers operational efficiency but also constrains the exploration of new ideas. Understanding the underlying reasons for the existence of these bureaucratic layers is imperative to devising effective mitigation strategies and instilling confidence in revised systems.

How can we expect our people to make tough decisions in wartime if we do not develop this trait in peacetime? The dichotomy between decision-making in peaceful contexts and the exigencies of conflict underscores the need to instil confidence in personnel to navigate complex situations effectively.

> *Delegate down beyond your comfort level;*
> *but wrap support systems around those decision-makers.*
> – Air Vice-Marshal Harvey Reynolds,
> Air and Space Power Conference 2024

In peace time, fear of scrutiny from the public, media, and government often leads to decision paralysis, particularly among lower-ranking individuals. To mitigate this, there is a pressing need to decentralise decision-making authority and empower personnel at all levels to make informed choices. By providing opportunities for individuals to grapple with tough decisions in non-crisis situations, organisations can better prepare them for the rigours of wartime decision-making.

Central to this approach is understanding the boundaries of innovation and knowing when to challenge conventional rules and norms. Rather than solely focusing on decisions related to firing or not firing, personnel should feel emboldened to challenge existing systems and propose alternative approaches to problem-solving. To foster a culture conducive to empowered decision-making and innovation, organisations can implement the following strategies:

- **Consult widely:** Encourage open dialogue and idea-sharing across all ranks to help foster an environment where diverse perspectives are valued and respected.
- **Explore diverse options:** Host regular creative sessions where unconventional ideas are welcomed and encouraged, providing a platform for military design thinking and innovation.
- **Create safe thinking space:** Implement "no penalty" policies for suggesting innovative, albeit risky, ideas to ensure that personnel feel supported in taking calculated risks and exploring new avenues for improvement.
- **Develop emerging decision-makers:** Empower this positive failure principle down to the professional military education schoolhouses through realistic scenario-based learning so that experience can be gained in a safe space.

6.5.6 *Promote Creative White Space*

The concept of *creative white space* entails the deliberate allocation of time, conscientiously scheduled into daily routines, for the purposes of contemplation, creativity, strategic planning, and intellectual enrichment. This idea exemplifies the practice of dedicating specified intervals for reading, reflection, and intellectual engagement. Engaging in both structured and unstructured activities, such as competitions, games, and social activities, along with allocated time for exploring workplace issues in more depth, should empower and inspire (Gallate et al., 2012). Furthermore, select positions should be established where creative white space is built into the role. Rotating staff through such cognitive incubator posts could help stimulate innovation in future positions.

Operationalising creative white space entails judiciously reprioritising and reallocating time away from bureaucratic tasks that offer minimal value to core warfighting functions. Historically, the military has designated specific periods, such as physical fitness time or socialising sessions, within the weekly schedule. By reassessing the necessity of these traditional time allocations, leaders can liberate precious hours for fostering creativity and strategic contemplation.

Embracing contemporary mediums of learning and intellectual stimulation presents an avenue for maximising creative white space. For instance, leaders often capitalise on commute time by incorporating educational podcasts into their daily routines. Leveraging technological advancements, such as audiobooks and digital resources, further amplifies opportunities for continuous learning and ideation. These points are also worth considering:

- **Reallocate time for creativity:** Reprioritise schedules by reducing low-value bureaucratic tasks and create designated intervals for contemplation, creativity, and strategic planning to foster innovation.
- **Utilise modern learning tools:** Incorporate contemporary mediums like podcasts, audiobooks, and digital resources into daily routines to maximise intellectual enrichment and continuous learning.

6.5.7 *Discourage Perfectionism*

The axiom "Perfection is the enemy of innovation" underscores the inherent tension between the quest for flawless solutions and the imperative to iterate and evolve over time. Embracing the principle of accepting the 80% solution serves as a catalyst for fostering innovation within military contexts. Rather than expending exhaustive efforts in the pursuit of absolute perfection, leaders are encouraged to adopt a pragmatic approach that prioritises rapid deployment and iterative refinement. By embracing the 80% solution, organisations can expedite the implementation of concepts or interventions, thereby facilitating the iterative process of improvement and adaptation based on real-world feedback and experience. However, this is subjective and prone to personality differences, as a considerable number of meticulous and fastidious individuals in the organisation may resist this approach. It is essential to balance these tendencies with the need for timely and effective innovation, thus ensuring the pursuit of perfection does not stifle progress and adaptability.

Central to this paradigm shift is a departure from the prevailing inclination to mitigate risk by waiting for solutions to be 100% proven and tested before deployment. Acknowledging the inherent uncertainty of warfare, leaders are urged to embrace calculated risks and deploy solutions that may initially be deemed only 70% viable (albeit the actual value is arbitrary). As General Patton famously quipped, "A good plan, violently executed now, is better than a perfect plan next week." This willingness to accept a degree of imperfection enables organisations to swiftly deploy innovations into the battlefield, where they can be refined and optimised through practical application and experiential learning. This ethos extends beyond tactical capabilities to encompass operational planning and doctrinal development.

The circular argument that inhibits the inclusion of unproven concepts in doctrine is challenged by the acceptance of the so-called 80% solution. By recognising the value of deploying nascent ideas into the battlefield for validation, organisations can break free from the doctrinal inertia that perpetuates reliance on proven solutions. By deploying solutions at an earlier stage of development, organisations can streamline support and training efforts, minimising the resources expended on perfecting solutions before deployment. Three concepts worth promoting:

- **Embrace the 80% solution:** Prioritise rapid deployment and iterative refinement over exhaustive pursuit of perfection to foster innovation and adapt solutions based on real-world feedback.
- **Balance perfectionist tendencies:** Recognise and address the unhelpful inertia created by meticulous and fastidious individuals to ensure their pursuit of perfection does not stifle time-sensitive innovation.
- **Encourage calculated risks:** Shift from waiting for 100% proven solutions to accepting and deploying 70%–80% viable solutions, enabling swift deployment and refinement through practical application and experiential learning.

6.5.8   *Listen to the Subject Matter Experts*

Military organisations frequently commission reviews, reports, and audits, often relying on subject matter experts to provide valuable insights and recommendations. However, the mere act of commissioning these assessments is sometimes perceived as sufficient action without the follow-through needed to implement the expert advice. Effective leadership entails the courage to act upon these recommendations and leverage the expertise provided by subject matter experts. Unfortunately, many of these reports end up collecting dust on shelves or are dismissed by decision-makers due to the perceived difficulty of implementing the suggested changes. Often, if the experts do not provide the desired conclusions, their recommendations are disregarded.

To overcome this directionless challenge of remediation, military leaders must demonstrate the willingness to make tough decisions and establish partnerships built on trust with external entities, such as think tanks and research institutions. By building networks with these organisations, military leaders can access specialised subject matter advice and collaborate on joint projects. Key strategies for leveraging subject matter expertise include:

- **Network externally:** Establishing partnerships with civilian research and industry entities and think tanks to access specialised knowledge and expertise.
- **Collaborate:** Engaging in joint research projects on topics such as technology, strategy, and logistics to foster collaboration between military and civilian experts.
- **Incentivise research:** Encouraging personnel to participate in external research initiatives, providing opportunities for professional development and knowledge exchange.

6.5.9   Grow Expert Facilitators

To fully leverage the cognitive diversity within a group, expert facilitators are essential components of the innovation ecosystem. Various collaborative design methodologies and approaches, such as co-design and military design thinking, have been discussed extensively throughout this book as initial steps towards addressing complex challenges.

The application of military design thinking and collaborative design approaches has proliferated across the services, often as a consulting service, due to the inability to generate such expertise in-house. These approaches can help to harness transformative and innovative capabilities. However, the adoption of these methodologies has led to the absence of professional skill sets typically possessed by expert, unbiased facilitators.

For example, military design facilitation practitioners are not all equal; they vary significantly across design and non-design disciplines with various expertise and experience backgrounds. This diversity underscores the complex and multifaceted nature of design facilitation practice (Mosely et al., 2021), with the capabilities, practice, and expertise of design practitioners closely intertwined and influenced by the immediate context in which the practice occurs (Wrigley & Mosely, 2022). The impact of these dynamics on outcomes is significant, yet shortcomings are often erroneously attributed to the field of design itself. Therefore, it is imperative to ensure the expertise and experience of the facilitators to enable and draw out creative collaboration with users and stakeholders to ensure effective and innovative solutions. Key ideas include:

- **Use effective facilitators:** Employ professionally trained and experienced expert design facilitators to enhance creative collaboration with users and stakeholders, ensuring effective and innovative outcomes.
- **Ensure diversity of facilitators:** Recognise and address the diverse backgrounds and expertise of design facilitation practitioners, as their varied experiences significantly influence the outcomes and effectiveness of design initiatives.

6.5.10   *Increase Epistemic Diversity*

Expanding the diversity of thought within the military has many benefits, but two stand out. The first is that the military should represent a cross-section of the nation they are defending, but the second main reason is that such diversity enhances innovation mindsets. While domestic recruiting helps support the first reason, opportunities to recruit internationally can also have benefits. By tapping into expertise from various backgrounds, this approach not only enriches the organisation with diverse perspectives but also enhances personnel development by exposing individuals to different working environments and best practices. This advocates for promoting cross-disciplinary learning to break down siloed work practices and operational rhythms within the military. Strategies include facilitating cross-training programs to broaden skill sets, encouraging participation in joint exercises with different military branches or civilian sectors, and establishing mentorship programs that pair individuals from diverse backgrounds. These initiatives aim to cultivate a more adaptable and versatile workforce while fostering knowledge exchange and professional growth within the organisation. Two key recommendations include:

- **Promote cross-disciplinary learning:** Implement cross-training programs and encourage participation in joint exercises with various military branches and civilian sectors to break down siloed practices and broaden skill sets.
- **Enhance recruitment diversity:** Expand recruiting efforts domestically and internationally to enrich the organisation with diverse perspectives, promoting innovation and exposing personnel to different working environments and best practices.

6.5.11   *Encourage Experimentation as Learning*

Encouraging experimentation and embracing failure as a learning opportunity is paramount for fostering innovation within military organisations. But it goes beyond this simple step – to cement learning, there needs to be facilitated reflection. To achieve this, leaders must create safe spaces where personnel feel empowered to test new ideas without the fear of repercussions.

Conducting after-action reviews that focus on learning from both failures and successes is essential for extracting valuable insights and improving future endeavours. Additionally, cultivating an environment where constructive feedback is encouraged and normalised facilitates continuous improvement and innovation.

It is imperative to reassess existing processes to ensure they meet the requirements for fostering innovation effectively. Approximately 75% of innovative solutions stem from informal interactions, highlighting the importance of informal settings such as drone games, competitions, and other unstructured activities. These platforms provide opportunities for bricolage, which is conducive to creative thinking and experimentation.

Investing in continuous education and training is another critical aspect of nurturing innovation within military organisations. Providing access to ongoing education in emerging technologies and methodologies, sponsoring participation in external workshops, conferences, and courses, and regularly updating training curriculums to include elements of creative thinking and problem-solving are essential strategies in this regard.

Forging industry links, such as shipbuilding design workshops, facilitates the cross-pollination of ideas and best practices between military and civilian sectors, consequently enriching the innovation ecosystem within the organisation. By embracing design thinking methodologies and leveraging industry collaborations, military organisations can foster a culture of innovation and adaptability, ensuring readiness for future challenges and opportunities.

By actively pursuing the following action items, leaders can cultivate an environment conducive to innovation, where catalysts for change are supported, and blockers are systematically addressed. This proactive approach is essential for maintaining a dynamic and effective military force in an increasingly complex and technologically advanced world.

- **Create safe spaces for experimentation:** Empower personnel to test new ideas without fear of repercussions, using after-action reviews to learn from both failures and successes.
- **Foster informal innovation and continuous education:** Encourage informal settings for creative thinking, such as competitions and unstructured activities, while investing in continuous education and industry collaborations to enhance innovation and adaptability.

6.5.12 *Increase and Promote More "Red Teaming"*

Sun Tzu's timeless adage "Know yourself, but know your enemy better" underscores the critical importance of empathy. In contemporary military strategy, the concept of "red teaming" emerges as a powerful tool to achieve this understanding and maintain strategic advantage. This concept is not just for the intelligence community – or, to put it another way, everyone is an intelligence analyst. Empathy is about thinking like the other influencers of the system. Anticipating their next move only comes from *thinking like them*.

Red teaming involves the deliberate simulation of adversary actions through frequent attacks on one's own systems, processes, and national security vulnerabilities. By engaging skilled adversaries in these exercises, military forces can unearth vulnerabilities that may otherwise remain undetected, allowing for proactive remediation before exploitation occurs. Despite the inherent reluctance to expose weaknesses within one's own systems, the actionable insights gleaned from red teaming activities are paramount.

However, the efficacy of red teaming often falters in practice, with valuable lessons and warning signs frequently relegated to the sidelines, unaddressed. To realise its full potential, red teaming must undergo a paradigm shift. It demands a reimagined approach – one that transcends mere simulation exercises – to foster a culture of continuous learning and improvement. Only through proactive engagement with the insights garnered from red teaming can military forces truly fortify their defences and maintain their competitive edge in an ever-evolving threat landscape. The following suggestions are worth consideration:

- **Implement regular red teaming exercises:** Conduct frequent simulations to identify and address vulnerabilities in systems and processes to enhance strategic readiness.
- **Cultivate a culture of continuous learning:** Ensure insights from red teaming exercises are actively reviewed and integrated into continuous improvement efforts to foster proactive defence strategies.

6.5.13   *Use an Evolutionary Approach*

Failure is simply the opportunity to begin again, this time more intelligently.
– Henry Ford

In alignment with Henry Ford's wisdom, we advocate for the military's adoption of an evolutionary approach to integrating new technologies and warfare concepts. Embracing agile methodologies in development and deployment offers a strategic advantage, enabling the military to navigate the complexities of modern warfare while mitigating the risks associated with rapid technological advancement. By fostering adaptability, this approach ensures that military doctrine and culture evolve alongside emerging technologies, aligning organisational changes with strategic objectives and operational requirements. An evolutionary strategy prioritises risk mitigation by allowing for incremental changes, minimising the potential for unintended consequences like excessive micromanagement or heightened vulnerability. This proactive stance facilitates thorough testing and evaluation of new technologies before widespread implementation, enhancing operational efficiency and effectiveness.

To implement this recommendation, the military should establish cross-functional agile development teams tasked with collaboratively refining new technologies and concepts. These teams would prioritise feedback and adaptation, leveraging open communication channels to gather insights from end users and stakeholders. By fostering a culture of continuous improvement, the military can encourage innovation and empower personnel to contribute towards driving technological advancements and operational excellence. Additionally, investing in modern infrastructure and logistical support for agile initiatives is crucial to ensuring successful implementation. Through these measures, the military can capitalise on an evolutionary approach to innovation, effectively integrating new technologies and concepts of warfare while maintaining operational readiness in a dynamic security environment. It is worth considering the following:

- **Adopt agile methodologies:** Establish cross-functional agile development teams to iteratively refine and implement new technologies and concepts, prioritising feedback and adaptation to enhance operational efficiency.
- **Foster a culture of continuous improvement:** Encourage innovation and empower personnel by promoting open communication, investing in modern infrastructure, and supporting agile initiatives to maintain operational readiness and drive technological advancements.

6.5.14 *Incentivise the Bold*

In the military context, fostering a culture that encourages boldness and innovation is paramount for staying ahead in an ever-evolving landscape of threats and challenges. Much like intense experiences in life, such as sailing a yacht or performing on stage, military activities that unite a team through common hardship help to form a bond that can never be understood by those who have not experienced it.

For genuine risk-takers in the military, success and failure are not seen as opposites but rather as vivid experiences that evoke strong emotions. Those who take bold chances understand that complacency, not failure, is the true adversary. Success and failure alike stem from action, creativity, and daring and are essential components of the journey towards innovation and excellence.

By embracing a mindset that views failure as a learning opportunity rather than a setback, the military can cultivate a culture of experimentation, growth, and resilience. Just as in the corporate world, where failure is often seen as a stepping stone to success, the military can reframe setbacks as valuable learning experiences on the path towards achieving strategic objectives. This failure-friendly attitude is not new; it has been embraced by pioneering figures throughout history, including renowned inventors like Thomas Edison and the Wright Brothers. By embracing failure as an integral part of the innovation process, the military can harness the power of adversity to drive continuous improvement and maintain a competitive edge in an increasingly complex security environment. Two suggestions include:

- **Encourage boldness and innovation:** Foster a culture that rewards risk-taking and views failure as a valuable learning opportunity, promoting experimentation and continuous improvement.
- **Unite teams through shared hardships:** Utilise challenging experiences to strengthen team bonds and resilience, enhancing collective creativity and daring in pursuit of strategic objectives.

6.6   Final Thoughts

In our exploration of redefining success, we have emphasised the imperative need for change, adaptation, and innovation. The current military landscape demands a departure from traditional thinking and the cultivation of a new perspective that thrives in complexity.

In the quest to redefine success, we envision a modern military that not only embraces change but actively seeks it. Success should no longer be confined to traditional measures but should encompass adaptability and creativity as core tenets. We look forward to a future where diversity, innovation, and adaptability are embraced, enabling military forces to confidently and purposefully tackle the challenges of the 21st century.

This book has set the tone for the urgent transformation required; however, the real challenge lies in inspiring others to take meaningful action to make it a reality. Among those in positions of influence, who will possess the courage to transcend mere symbolic gestures and take substantive action? Such courage is inherently uncomfortable and demands a willingness to take risks. Does this necessitate the emergence of a true maverick leader to pioneer the way forward? One such maverick involved in this research believed that our military today exemplifies the principle:

True innovation occurs at the intersection of the impossible and the necessary.

Perhaps this is precisely the juncture at which the Free World currently finds itself, demanding innovation at this nexus point in history – but on which side will we land?"

References

Aalbers, R., Dolfsma, W., & Koppius, O. (2013). Individual connectedness in innovation networks: On the role of individual motivation. *Research Policy, 42*(3), 624–634. https://doi.org/10.1016/j.respol.2012.10.007

Abdukakharovna, A. M. (2020). The importance of military traditions in improving the professional training of military servants. *The American Journal of Social Science and Education Innovations, 2*(9), 647–651. https://doi.org/10.37547/tajssei/Volume02Issue09-99

Ackoff, R. L. (1981). On the use of models in corporate planning. *Strategic Management Journal, 2*(4), 353–359. https://doi.org/10.1002/smj.4250020404

Ackoff, R. L. (1994). Systems thinking and thinking systems. *System Dynamics Review, 10*(2–3), 175–188. https://doi.org/10.1002/sdr.4260100206

Ackoff, R. L. (2015). *A lifetime of systems thinking.* The Systems Thinker. https://thesystemsthinker.com/a-lifetime-of-systems-thinking/

Ackoff, R. L., Magidson, J., & Addison, H. J. (2006). *Idealized design: Creating an organization's future.* Pearson Education.

Allison, R. (2020). Russian revisionism, legal discourse and the 'rules-based' international order. *Europe-Asia Studies, 72*(6), 976–995. https://doi.org/10.1080/09668136.2020.1773406

Amyx, S. (2019, September 3). How constraints help or inhibit innovation. *Forbes.* www.forbes.com/sites/forbesnycouncil/2019/09/03/how-constraints-help-or-inhibit-innovation/?sh=6c14a30a5b82

Ancker, C. J., III, & Flynn, M. (2010). Field manual 5–0: Exercising command and control in an era of persistent conflict-FM 5–0 represents a significant evolution in army doctrine for focusing on complex environments. *Military Review, 90*(2), 13.

Anderson, E. G., Jr., Lim, S. Y., & Joglekar, N. (2017). Are more frequent releases always better? Dynamics of pivoting, scaling, and the minimum viable product. In T. X. Bui & R. Sprague, Jr. (Eds.), *Proceedings of the 50th Hawaii international conference on system sciences* (pp. 5849–5858). University of Hawaii at Manoa. https://doi.org/10.24251/HICSS.2017.705

Antrobus, S., & West, H. (2022). 'This is all very academic': Critical thinking in professional military education. *The RUSI Journal, 167*(3), 78–86. https://doi.org/10.1080/03071847.2022.2112521

Artto, K., Martinsuo, M., Dietrich, P., & Kujala, J. (2008). Project strategy: Strategy types and their contents in innovation projects. *International*

Journal of Managing Projects in Business, 1(1), 49–70. https://doi.org/10.1108/17538370810846414

Assink, M. (2006). Inhibitors of disruptive innovation capability: A conceptual model. *European Journal of Innovation Management, 9*(2), 215–233. https://doi.org/10.1108/14601060610663587

Augier, M., McNab, R., Guo, J., & Karber, P. (2017). Defense spending and economic growth: Evidence from China, 1952–2012. *Defence and Peace Economics, 28*(1), 65–90. https://doi.org/10.1080/10242694.2015.1099204

Banach, S. J. (2009, March–April). Educating by design: Preparing leaders for a complex world. *Military Review, 89*, 96–104.

Banach, S. J., & Ryan, A. (2009, March–April). The art of design: A design methodology. *Military Review, 89*, 105–115.

Barker, G. (2014, March 6). Army leaders crusade for $10bn plus vehicle upgrades. *Australian Financial Review*.

Beaulieu-Brossard, P., & Dufort, P. (2017). Introduction: Revolution in military epistemology. *Journal of Military and Strategic Studies, 17*(4), 1–20.

Bell, C. R., & Patterson, J. R. (2005). Command presence: Animate and engage people. *Leadership Excellence, 22*(12), 11.

Bentley, B. (2005). *Professional ideology and the profession of arms in Canada*. Canadian Institute of Strategic Studies.

Bergkvist, L., & Karlsson, J. (2019). Bridging the gap – From great ideas to realized innovations. In P. Kristensson, P. Magnusson, & L. Witell (Eds.), *Service innovation for sustainable business: Stimulating, realizing and capturing the value from service innovation* (pp. 225–252). World Scientific. https://doi.org/10.1142/9789813273382_0012

Bhaduri, S., & Kumar, H. (2011). Extrinsic and intrinsic motivations to innovate: Tracing the motivation of 'grassroot' innovators in India. *Mind & Society, 10*(1), 27–55. https://doi.org/10.1007/s11299-010-0081-2

Black, C. N., Newton, R. D., Nobles, M. A., & Ellis, D. C. (2018). U.S. special operations command's future, by design. *Joint Force Quarterly, 90*, 42–49.

Blomme, M. E. (2015). On theory: War and warfare reconsidered. *Army War College Review, 1*(1), 24–41.

Boyd, J. R. (2018). *A discourse on winning and losing* (Vol. 400). Air University Press.

Bragdon, T. (2017, September 7). When data gives the wrong solution. *Pitch + Persuade*. www.trevorbragdon.com/p/when-data-gives-the-wrong-solution

Brenk, S. (2020). Open business model innovation – The impact of breadth, depth, and freedom of collaboration. *Academy of Management Proceedings, 1*. https://doi.org/10.5465/AMBPP.2020.19

Bruton, D. (2011). Learning creativity and design for innovation. *International Journal of Technology and Design Education, 21*(3), 321–333. https://doi.org/10.1007/s10798-010-9122-8

Buchanan, R. (1992). Wicked problems in design thinking. *Design Issues, 8*(2), 5–21. https://doi.org/10.2307/1511637

Burk, J. (1999). Military culture. In J. E. Turpin & L. R. Kurtz (Eds.), *Encyclopedia of violence, peace, and conflict* (Vol. 2, pp. 447–461). Academic Press.

Cardon, E. C., & Leonard, S. (2010, March–April). Unleashing design: Planning and the art of battle command. *Military Review, 90*, 2–12.

Carr, A. (2024). Strategy as problem-solving. *Parameters*, *54*(1), 10. https://doi.org/10.55540/0031-1723.3276

Chaleff, I. (2010). *The courageous follower: Standing up to & for our leaders* (ReadHowYouWant Ed.). Accessible Publishing Systems.

Chamorro-Premuzic, T. (2013, August 22). Why do so many incompetent men become leaders? *Harvard Business Review*.

Chandy, R. K., & Tellis, G. J. (2000). The incumbent's curse? Incumbency, size, and radical product innovation. *Journal of Marketing*, *64*(3), 1–17. https://doi.org/10.1509/jmkg.64.3.1.18033

Cheng, X., Fu, S., de Vreede, T., de Vreede, G.-J., Seeber, I., Maier, R., & Weber, B. (2020). Idea convergence quality in open innovation crowdsourcing: A cognitive load perspective. *Journal of Management Information Systems*, *37*(2), 349–376. https://doi.org/10.1080/07421222.2020.1759344

Chung, Y. W., & Moon, H. K. (2011). The moderating effects of collectivistic orientation on psychological ownership and constructive deviant behavior. *International Journal of Business and Management*, *6*(12), 65–77. https://doi.org/10.5539/ijbm.v6n12p65

Churchman, C. W. (1971). *The design of inquiring systems: Some basic concepts of systems and organization*. Basic Books.

Clarke, M., Shah, A., & Sharma, U. (2011). Systematic review of studies on telemonitoring of patients with congestive heart failure: A meta-analysis. *Journal of Telemedicine and Telecare*, *17*(1), 7–14. https://doi.org/10.1258/jtt.2010.100113

Collins, J., & Mills, G. (2019). Digital age education: Preparing warriors for hybrid conflict at Air Force CyberWorx. In J. Ridolfo & W. Hart-Davidson (Eds.), *Rhet ops: Rhetoric and information warfare* (pp. 183–199). University of Pittsburgh Press. https://doi.org/10.2307/j.ctvqc6hmj.16

Cooney, S. M. (2012). *Re-introducing conceptual and detailed planning: Differentiating between decision making and problem identification*. United States Army Command and General Staff College.

Covey, S. R. (1991) *The seven habits of highly effective people*. Covey Leadership Center.

Dapra, R. A., Zarrillo, D. L., Carlson, T. K., & Teevan, R. C. (1985). Fear of failure and indices of leadership utilized in the training of ROTC cadets. *Psychological Reports*, *56*(1), 27–30. https://doi.org/10.2466/pr0.1985.56.1.27

de Czege, H. W. (2009, January–February). Systemic operational design: Learning and adapting in complex missions. *Military Review*, *89*(1), 2–12.

De Spiegeleire, S., Sweijs, T., Wijninga, P., & van Esch, J. (with Galdiga, J. H., Hsu, W.-C., & Komrij, F.). (2014). *Designing future stabilization efforts*. The Hague Centre for Strategic Studies.

den Besten, A. (2015). *Overconfidence through group composition: The effect of group composition on overconfidence* [Master's thesis, Erasmus University]. Erasmus University Thesis Repository. http://hdl.handle.net/2105/30228

DePaul, G. A. (2022). *Nine practices of 21st century leadership: A guide for inspiring creativity, innovation, and engagement* (2nd ed.). CRC Press. https://doi.org/10.4324/9781003273448

Dienel, G. A. (2019). Brain glucose metabolism: Integration of energetics with function. *Physiological Reviews*, *99*(1), 949–1045. https://doi.org/10.1152/physrev.00062.2017

Dufort, P. (2017). Carl von Clausewitz and the invention of the conservative nation-state: Retrieving instrumental reflexivity in the strategic tradition. *Journal of Military and Strategic Studies*, *17*(4), 209–236.

Duguid, M. M., & Goncalo, J. A. (2015). Squeezed in the middle: The middle status trade creativity for focus. *Journal of Personality and Social Psychology*, *109*(4), 589–603. https://doi.org/10.1037/a0039569

Dunivin, K. O. (1994). Military culture: Change and continuity. *Armed Forces & Society*, *20*(4), 531–547. https://doi.org/10.1177/0095327X9402000403

Dunk, G., & Kruger, J. (2023, June 6). Shaking Australia's 'state vs state' attitude to innovation. *The Strategist*. www.aspistrategist.org.au/shaking-australias-state-vs-state-attitude-to-innovation/

Dunning, D. (2011). The Dunning–Kruger effect: On being ignorant of one's own ignorance. In *Advances in experimental social psychology* (Vol. 44, pp. 247–296). Academic Press.

Edwards, G. (2022, June 17). Is Clausewitz dead? The problem with centre of gravity. *Wavell Room*. https://wavellroom.com/2022/06/17/is-clausewitz-dead-complex-adaptive-systems-operations-planning/

Ellemers, N., & Haslam, S. A. (2012). Social identity theory. In P. A. M. Van Lange, A. W. Kruglanski, & E. T. Higgins (Eds.), *The handbook of theories of social psychology* (Vol. 2, pp. 379–398). SAGE Publications. https://doi.org/10.4135/9781446249222

Ellenberg, J. (2014). *How not to be wrong: The hidden maths of everyday life*. Penguin UK.

Ellis-Smith, J. (2018). The network model: Options for a future Australian Army. *Australian Army Journal*, *14*(1), 19–39.

Ervin, D. D. (2020). *Mission command and the intelligence commander – In chaos lies opportunity: A model for creating belief, capability, and driving intelligence operations*. Air War College, Air University.

Euler, L. (1741). Solutio problematis ad geometriam situs pertinentis [The solution of a problem relating to the geometry of position]. *Commentarii Academiae Scientiarum, Petropolitanae*, *8*, 128–140.

Evans, M. (2008). The closing of the Australian military mind: The ADF and operational art. *Security Challenges*, *4*(2), 105–131. www.jstor.org/stable/26459144

Fan, H.-L., Huang, M.-H., & Chen, D.-Z. (2019). Do funding sources matter?: The impact of university-industry collaboration funding sources on innovation performance of universities. *Technology Analysis & Strategic Management*, *31*(11), 1368–1380. https://doi.org/10.1080/09537325.2019.1614158

Finkel, M. (2019). Conservatism by choice (stability) – A necessary complement to innovation and adaptation in force design. *Defence Studies*, *19*(4), 392–409. https://doi.org/10.1080/14702436.2019.1686359

Fischer, C., Malycha, C. P., & Schafmann, E. (2019). The influence of intrinsic motivation and synergistic extrinsic motivators on creativity and innovation. *Frontiers in Psychology*, *10*, 137. https://doi.org/10.3389/fpsyg.2019.00137

Frank, C., Sink, C., Mynatt, L., Rogers, R., & Rappazzo, A. (1996). Surviving the "valley of death": A comparative analysis. *The Journal of Technology Transfer, 21*(1–2), 61–69. https://doi.org/10.1007/BF02220308

Freedman, L. (2017). *The future of war: A history.* PublicAffairs.

Frumin, A., Moss, T., & Ellis, D. C. (2018). The state of the art in contemporary CWMD thinking. *PRISM, 7*(3), 68–83.

Gallate, J., Wong, C., Ellwood, S., Roring, R. W., & Snyder, A. (2012). Creative people use nonconscious processes to their advantage. *Creativity Research Journal, 24*(2–3), 146–151.

Gardiner, E., & Jackson, C. J. (2012). Workplace mavericks: How personality and risk-taking propensity predicts maverickism. *British Journal of Psychology, 103*(4), 497–519. https://doi.org/10.1111/j.2044-8295.2011.02090.x

Gladwell, M. (2021). *The bomber mafia: A story set in war.* Penguin Books.

Goldwater-Nichols Department of Defense Reorganization Act of 1986 (1986). www.congress.gov/bill/99th-congress/house-bill/3622

Graicer, O. (2017). Self disruption: Seizing the high ground of systemic operational design (SOD). *Journal of Military and Strategic Studies, 17*(4), 21–37.

Grant, A. M. (2013). Rocking the boat but keeping it steady: The role of emotion regulation in employee voice. *Academy of Management Journal, 56*(6), 1703–1723. https://doi.org/10.5465/amj.2011.0035

Greene, R. (2020). Research for practitioners: Cognitive bias. *Compensation & Benefits Review, 52*(1), 5–7. https://doi.org/10.1177/0886368719892176

Groysberg, B., Lee, J., Price, J., & Cheng, J. Y.-J. (2018, January–February). The leader's guide to corporate culture: How to manage the eight critical elements of organizational life. *Harvard Business Review, 96*(1), 44–52.

Haesevoets, T., De Cremer, D., Hirst, G., De Schutter, L., Stouten, J., van Dijke, M., & Van Hiel, A. (2022). The effect of decisional leader procrastination on employee innovation: Investigating the moderating role of employees' resistance to change. *Journal of Leadership & Organizational Studies, 29*(1), 131–146. https://doi.org/10.1177/15480518211044166

Harężlak, P., & Rosa, R. (2019). War studies university attempt to design thinking. *Zeszyty Naukowe Akademii Sztuki Wojennej, 1*(114), 75–87.

Heltberg, T. (2022). Military officers and the design gaze: A study of how Danish military officers experience challenges and possibilities when seeking to apply design approaches in their everyday work. *Journal of Military and Strategic Studies, 21*(3), 35–56. https://jmss.org/article/view/74852/56005

Heltberg, T., Krogh, A. H., & Kyne, K. (2023). Military design in European defense forces: An evolving niche. In A. M. Sookermany (Ed.), *Handbook of military sciences* (pp. 1–18). Springer. https://doi.org/10.1007/978-3-030-02866-4_113-1

Herrero, L. (2008). *Viral change: The alternative to slow, painful and unsuccessful management of change in organisations* (2nd ed.). Meetingminds Publishing.

Hosseinzadeh, M., & Yoosefi, N. (2022). *Problem-solving strategies of the introvert and extrovert translation students* [Paper presentation]. 17th International TELLSI Conference, Islamic Azad University of Tabriz, Iran.

Howell, J. M., Shea, C. M., & Higgins, C. A. (2005). Champions of product innovations: Defining, developing, and validating a measure of champion behavior. *Journal of Business Venturing, 20*(5), 641–661. https://doi.org/10.1016/j.jbusvent.2004.06.001

Jackson, A. P. (2019, February 26–28). *Civilian and military design thinking: A comparative historical and paradigmatic analysis, and its implications for military designers* [Paper presentation]. Innovation Methodologies for Defence Challenges (IMDC) 2019 Conference, Lancaster University, United Kingdom.

Jackson, D., & Humble, J. (1994). Middle managers: New purpose, new directions. *Journal of Management Development, 13*(3), 15–21. https://doi.org/10.1108/02621719410051966

Jalonen, H. (2012). The uncertainty of innovation: A systematic review of the literature. *Journal of Management Research, 4*(1), 1–52.

Jans, N., & Cullens, J. (2010). Analysis of ethical conduct within the Australian Defence Forces. In J. Stouffer & S. Seiler (Eds.), *Military ethics: International perspectives* (pp. 163–184). Canadian Defence Academy Press.

Jans, N., & Frazer-Jans, J. (2004). Career development, job rotation, and professional performance. *Armed Forces & Society, 30*(2), 255–277. https://doi.org/10.1177/0095327X0403000206

Jensen, A. R. (1998). *The g factor: The science of mental ability.* Praeger Publishers.

Johansen, R. B., Laberg, J. C., & Martinussen, M. (2013). Measuring military identity: Scale development and psychometric evaluations. *Social Behavior and Personality: An International Journal, 41*(5), 861–880. https://doi.org/10.2224/sbp.2013.41.5.861

Jordan, R. (2019). *Maverickism: Round pegs in square holes: The who, what and why of mavericks in organisations* [Doctoral dissertation, The University of Queensland]. UQ eSpace. https://doi.org/10.14264/uql.2019.795

Jordan, R. (2022a). Challenge accepted: How to harness your Top Gun talent for greater workplace benefits. *Momentum: The Business School Magazine.* https://stories.uq.edu.au/momentum-magazine/challenge-accepted-workplace-mavericks/index.html

Jordan, R. (2022b). Planning to start a new job? Learn how workplace 'mavericks' influence early on and retain uniqueness + tips for your first 90 days in a new gig. *Momentum: The Business School Magazine.* https://stories-uq-edu-au.cdn.ampproject.org/c/s/stories.uq.edu.au/momentum-magazine/learn-how-to-workplace-mavericks-influence/amp-index.html

Jordan, R., Fitzsimmons, T. W., & Callan, V. J. (2021). Fear not your mavericks! Their bounded non-conformity and positive deviance helps organizations drive change and innovation. In T. J. Andersen (Ed.), *Strategic responses for a sustainable future: New research in international management* (pp. 123–146). Emerald Publishing. https://doi.org/10.1108/978-1-80071-929-320214006

Jordan, R., Fitzsimmons, T. W., & Callan, V. J. (2023). Positively deviant: New evidence for the beneficial capital of maverickism to organizations. *Group & Organization Management, 48*(5), 1254–1305. https://doi.org/10.1177/10596011221102297

Jungdahl, A. M., & Macdonald, J. M. (2015). Innovation inhibitors in war: Overcoming obstacles in the pursuit of military effectiveness. *Journal of Strategic Studies*, *38*(4), 467–499. https://doi.org/10.1080/01402390.2014.917628

Kabutaulaka, T. T. (2010). Milking the dragon in Solomon Islands. In T. Wesley-Smith & E. A. Porter (Eds.), *China in Oceania: Reshaping the Pacific* (pp. 133–150). Berghahn Books. https://doi.org/10.2307/j.ctt1btbxk9.15

Kahneman, D. (2011). *Thinking, fast and slow*. Farrar, Straus and Giroux.

Kalms, M. (2020, April 3). Cultural reform in middle army – Melting iron colonels. *The Cove*. https://cove.army.gov.au/article/cultural-reform-middle-army-melting-iron-colonels

Kalms, M., & Sayers, J. (2020, April 3). Cultural reform in middle army – Melting iron colonels. *The Cove*. https://cove.army.gov.au/article/cultural-reform-middle-army-melting-iron-colonels

Kardos, M., & Dexter, P. (2017). *A simple handbook for non-traditional red teaming*. Australian Government Department of Defence.

Karjalainen, A., Alha, K., & Jutila, S. (2006). *Give me time to think: Determining student workload in higher education*. Oulu University Press.

Kelley, D., & Lee, H. (2010). Managing innovation champions: The impact of project characteristics on the direct manager role. *Journal of Product Innovation Management*, *27*(7), 1007–1019. https://doi.org/10.1111/j.1540-5885.2010.00767.x

Kern, T. (1999). *Darker shades of blue: The rogue pilot*. McGraw-Hill.

Kern, T. (2009). *Blue threat: Why to err is inhuman: How to wage and win the battle within*. Pygmy Books.

Kern, T. (2011). *Going pro: The deliberate practice of professionalism*. Pygmy Books.

Khalil, R., Godde, B., & Karim, A. A. (2019). The link between creativity, cognition, and creative drives and underlying neural mechanisms. *Frontiers in Neural Circuits*, *13*, 18. https://doi.org/10.3389/fncir.2019.00018

Kilcullen, D., Neuhaus, S., & Bridgewater, F. (2001). Military medical ethics: Issues for 21st century operations. *Australian Defence Force Journal*, *2001*(151), 49–58.

Kim, D., & Lee, D. (2018). Impacts of metacognition on innovative behaviors: Focus on the mediating effects of entrepreneurship. *Journal of Open Innovation: Technology, Market, and Complexity*, *4*(2), 18. https://doi.org/10.3390/joitmc4020018

Koerner, M. M. (2014). Courage as identity work: Accounts of workplace courage. *Academy of Management Journal*, *57*(1), 63–93. https://doi.org/10.5465/amj.2010.0641

Kulzy, W. W., III. (2019). (Design) Thinking through strategic-level wargames for innovation solutions. *Phalanx*, *52*(2), 56–62.

LaManna, J. C., Salem, N., Puchowicz, M., Erokwu, B., Koppaka, S., Flask, C., & Lee, Z. (2009). Ketones suppress brain glucose consumption. In P. Liss, P. Hansell, D. F. Bruley, & D. K. Harrison (Eds.), *Advances in experimental medicine and biology: Vol. 645. Oxygen transport to tissue XXX* (pp. 301–306). Springer. https://doi.org/10.1007/978-0-387-85998-9_45

Larsen, G. D. (2011). Understanding the early stages of the innovation diffusion process: Awareness, influence and communication networks. *Construction Management and Economics, 29*(10), 987–1002. https://doi.org/10.1080/01446193.2011.619994

Laszlo, E. (1996). *The systems view of the world: A holistic vision for our time.* Hampton Press.

Leonard, S. (2022, August 9). Metrics gone wild: The military's dangerous obsession. *ClearanceJobs.* https://news.clearancejobs.com/2022/08/09/metrics-gone-wild-the-militarys-dangerous-obsession/

Leonard, D., & Sensiper, S. (1998). The role of tacit knowledge in group innovation. *California Management Review, 40*(3), 112–132. https://doi.org/10.2307/41165946

Lindsay, J. D., Perkins, C. A., & Karanjikar, M. R. (2009). *Conquering innovation fatigue: Overcoming the barriers to personal and corporate success.* John Wiley & Sons.

Luft, J., & Ingham, H. (1955). *The Johari window: A graphic model for interpersonal relations.* University of California Western Training Lab.

Mandel, D. R. (2020). The occasional maverick of analytic tradecraft. *Intelligence and National Security, 35*(3), 438–443. https://doi.org/10.1080/02684527.2020.1723830

Marchau, V. A. W. J., Walker, W. E., Bloemen, P. J. T. M., & Popper, S. W. (Eds.). (2019). *Decision making under deep uncertainty: From theory to practice.* Springer Nature. https://doi.org/10.1007/978-3-030-05252-2

Margetts, R. (2016). *Thinking about thinking: Insights for junior officers in the New Zealand Defence Force.* Victoria University of Wellington.

Marrin, S. (2007). Intelligence analysis: Structured methods or intuition? *American Intelligence Journal, 25*(1), 7–16. www.jstor.org/stable/44327067

Matheson, S. S. (2007). *Maverick in the sky: The aerial adventures of World War I flying ace Freddie McCall.* Frontenac House.

Maverick, L. A. (1942). The term "maverick," applied to unbranded cattle. *California Folklore Quarterly, 1*(1), 94–96. https://doi.org/10.2307/1495731

Maverick, M. A., & Maverick, G. M. (1921). *Memoirs of Mary A. Maverick arranged by Mary A. Maverick and her son Geo. Madison Maverick.* Alamo Printing Company.

Mazzara, J. N. (2011). Words mean things: Understand the terminology being used. *Marine Corps Gazette, 95*(6), 23–26.

McCaffrey, T. (2012). Innovation relies on the obscure: A key to overcoming the classic problem of functional fixedness. *Psychological Science, 23*(3), 215–218. https://doi.org/10.1177/0956797611429580

McMurry, R. N. (1974). *The maverick executive.* Amacom.

McPheat, S. (2022). *What is the Thomas Kilmann conflict management model? (with examples).* MTD Management Training Specialists. www.mtdtraining.com/blog/thomas-kilmann-conflict-management-model.htm

Mehrotra, S. (2022). The designer's gaze. *DOC.* www.doc.cc/articles/the-designers-gaze

Mellor, A., Mobilia, M., Redner, S., Rucklidge, A. M., & Ward, J. A. (2015). Influence of Luddism on innovation diffusion. *Physical Review E, 92*(1), 012806. https://doi.org/10.1103/PhysRevE.92.012806

Mergenthaler, P., Lindauer, U., Dienel, G. A., & Meisel, A. (2013). Sugar for the brain: The role of glucose in physiological and pathological brain function. *Trends in Neurosciences, 36*(10), 587–597. https://doi.org/10.1016/j.tins.2013.07.001

Milevski, L. (2012). Whence derives predictability in strategy? *Infinity Journal, 2*(4), 4–7.

Miller, C. R. (2012). *The 6,000 mile screwdriver is getting longer: Washington's strengthening grip*. United States Army War College.

Miron, E., Erez, M., & Naveh, E. (2004). Do personal characteristics and cultural values that promote innovation, quality, and efficiency compete or complement each other? *Journal of Organizational Behavior, 25*(2), 175–199. https://doi.org/10.1002/job.237

Mosely, G., Markauskaite, L., & Wrigley, C. (2021). Design facilitation: A critical review of conceptualisations and constructs. *Thinking Skills and Creativity, 42*, 100962. https://doi.org/10.1016/j.tsc.2021.100962

Mosely, G., Wright, N., & Wrigley, C. (2018). Facilitating design thinking: A comparison of design expertise. *Thinking Skills and Creativity, 27*, 177–189.

Moskos, C. C., Jr. (1977). From institution to occupation: Trends in military organization. *Armed Forces & Society, 4*(1), 41–50. https://doi.org/10.1177/0095327X7700400103

Moten, C., III, Newton, H., & Jackson, L. (2016). *TRAC innovative visualization techniques* (Report No. TRAC-M-TR-17–003). TRADOC Analysis Center. https://apps.dtic.mil/sti/citations/AD1025683

Murray, W., & Millett, A. R. (Eds.). (1998). *Military innovation in the interwar period*. Cambridge University Press.

Neubauer, A. C., & Fink, A. (2009). Intelligence and neural efficiency. *Neuroscience & Biobehavioral Reviews, 33*(7), 1004–1023. https://doi.org/10.1016/j.neubiorev.2009.04.001

Newdigate, M. T. (2014). *Integrating special operations forces operational design and joint doctrine*. United States Army Command and General Staff College. https://doi.org/10.21236/ADA614156

Noworol, C., Żarczyński, Z., Fafrowicz, M., & Marek, T. (2017). Impact of professional burnout on creativity and innovation. In W. B. Schaufeli, C. Maslach, & T. Marek (Eds.), *Professional burnout: Recent developments in theory and research* (pp. 163–175). Routledge. https://doi.org/10.4324/9781315227979-13

Nusem, E., Wrigley, C., & Matthews, J. (2017). Developing design capability in nonprofit organizations. *Design Issues, 33*(1), 61–75.

Odell, A. (2022). The Navy still punishes talented risk-takers. *Proceedings, 148*(5). US Naval Institute. www.usni.org/magazines/proceedings/2022/may/navy-still-punishes-talented-risk-takers

Orak, U., & Walker, M. H. (2021). Military service: A pathway to conformity or a school for deviance? *Crime & Delinquency, 67*(6–7), 1046–1069. https://doi.org/10.1177/0011128719850497

Owen, J. M. (2021). Two emerging international orders? China and the United States. *International Affairs, 97*(5), 1415–1431. https://doi.org/10.1093/ia/iiab111

Page, S. E. (2007). Making the difference: Applying a logic of diversity. *Academy of Management Perspectives, 21*(4), 6–20. https://doi.org/10.5465/amp.2007.27895335

Palazzo, A. (2012). The future war debate in Australia: Why has there not been one? Has the time for one now arrived? *Australian Defence Force Journal, 2012*(189), 5–20.

Parker, L. (2020, January 23). Plan Jericho and the internet of defence things. *Australian Defence Magazine.* www.australiandefence.com.au/defence/cyber-space/plan-jericho-and-the-internet-of-defence-things

Pearce, G., & Saul, P. (2008). Toward a framework for military health ethics. In F. Allhoff (Ed.), *Physicians at war: The dual-loyalties challenge* (pp. 75–88). Springer. https://doi.org/10.1007/978-1-4020-6912-3_5

Perez, C., Jr. (2011, March–April). A practical guide to design: A way to think about it and a way to do it. *Military Review, 91*, 41–51.

Peter, L. J., & Hull, R. (1969). *The Peter principle: Why things always go wrong.* William Morrow & Co.

Petty, D. J., & O'Byrne, S. (2024). Linearly implicit integration of vibrational master equation using automatic differentiation. *AIP Conference Proceedings, 2996*(1), 060009. AIP Publishing. https://doi.org/10.1063/5.0187375

Pidd, M. (1997). Tools for thinking – Modelling in management science. *Journal of the Operational Research Society, 48*(11), 1150. https://doi.org/10.1057/palgrave.jors.2600969

Preston, V., Huddleston, M. S., & Chiaramonte, M. V. (2019). *Air Force CyberWorx design event: Air force cyber talent management* (Air Force CyberWorx Report 19–001). https://apps.dtic.mil/sti/citations/AD1105336

Price, R., Matthews, J., & Wrigley, C. (2018). Three narrative techniques for engagement and action in design-led innovation. *She Ji: The Journal of Design, Economics, and Innovation, 4*(2), 186–201. https://doi.org/10.1016/j.sheji.2018.04.001

Proctor, P. (2011, March–April). Fighting to understand: A practical example of design at the battalion level. *Military Review, 91*, 69–78.

Rabelo, R. J., & Bernus, P. (2015). A holistic model of building innovation ecosystems. *IFAC-PapersOnLine, 48*(3), 2250–2257. https://doi.org/10.1016/j.ifacol.2015.06.423

Raska, M. (2015). *Military innovation in small states: Creating a reverse asymmetry.* Routledge. https://doi.org/10.4324/9781315766720

Read, A. (2000). Determinants of successful organisational innovation: A review of current research. *Journal of Management Practice, 3*(1), 95–119.

Reed, G., Bullis, C., Collins, R., & Paparone, C. (2004). Mapping the route of leadership education: Caution ahead. *Parameters, 34*(3), 46–60. https://doi.org/10.55540/0031-1723.2220

Rittel, H. W. J., & Webber, M. M. (1973). Dilemmas in a general theory of planning. *Policy Sciences, 4*(2), 155–169. https://doi.org/10.1007/BF01405730

Robinson, K. (2006, February). *Do schools kill creativity?* [Video]. TED Conferences. www.ted.com/talks/sir_ken_robinson_do_schools_kill_creativity

Rock, D. (2009). *Your brain at work: Strategies for overcoming distraction, regaining focus, and working smarter all day long.* HarperCollins US.

Rodriguez, C. A., Walton, T. C., & Chu, H. (2020). Putting the "FIL" into "DIME": Growing joint understanding of the instruments of power. *Joint Force Quarterly, 97*, 121–128.

Rosen, S. P. (1991). *Winning the next war: Innovation and the modern military*. Cornell University Press.

Rosenberg, A. (2001). Reductionism in a historical science. *Philosophy of Science, 68*(2), 135–163. https://doi.org/10.1086/392870

Rosser, J. B., Jr. (1999). On the complexities of complex economic dynamics. *Journal of Economic Perspectives, 13*(4), 169–192. https://doi.org/10.1257/jep.13.4.169

Ryan, A. J. (2011). Applications of complex systems to operational design. In H. Sayama, A. Minai, D. Braha, & Y. Bar-Yam (Eds.), *Unifying themes in complex systems: Vol. VIII. Proceedings of the eighth international conference on complex systems* (pp. 1252–1266). NECSI Knowledge Press. https://doi.org/10.1016/B978-0-444-52076-0.50024-9

Rynes, S. L., Gerhart, B., & Parks, L. (2005). Personnel psychology: Performance evaluation and pay for performance. *Annual Review of Psychology, 56*, 571–600. https://doi.org/10.1146/annurev.psych.56.091103.070254

Schmidtchen, D. (2001). Developing creativity and innovation through the practice of mission command. *Australian Defence Force Journal, 2001*(146), 11–17.

Schnaubelt, C. M. (2009). Complex operations and *interagency* operational art. *PRISM, 1*(1), 37–50.

Shane, S. (1995). Uncertainty avoidance and the preference for innovation championing roles. *Journal of International Business Studies, 26*(1), 47–68. https://doi.org/10.1057/palgrave.jibs.8490165

Sharpe, B., Hodgson, A., Leicester, G., Lyon, A., & Fazey, I. (2016). Three horizons: A pathways practice for transformation. *Ecology and Society, 21*(2). www.jstor.org/stable/26270405

Sibley, M. M. (1973). *George W. Brackenridge: Maverick philanthropist*. University of Texas Press.

Simons, M. V. (2008). Professional ideology in the New Zealand Defence Force. In J. Stouffer & J. C. Wright (Eds.), *Professional ideology and development: International perspectives* (pp. 65–95). Canadian Defence Academy Press.

Simons, M. V. (2009). *Holistic professional military development: Growing strategic artists* [Doctoral dissertation]. Massey University. http://hdl.handle.net/10179/1434

Snowden, D. J. (2024). The uncertainty matrices. *The Cynefin Co.* https://tinyurl.com/muteud43

Snowden, D. J., & Boone, M. E. (2007, November). A leader's framework for decision making. *Harvard Business Review, 85*(11), 68–76.

Solbach, J., Möller, K., & Wirnsperger, F. (2022, April 20). Break the link between pay and motivation. *MIT Sloan Management Review.* https://sloanreview.mit.edu/article/break-the-link-between-pay-and-motivation/

Soni, P. (2021, January 12). For creative solutions, slow and deliberate thinking is the key. *Live Mint.* https://lifestyle.livemint.com/news/talking-point/for-creative-solutions-slow-and-deliberate-thinking-is-the-key-111641647009377.html

Sora, M. (2022). *A security agreement between China and Solomon Islands could impact stability in the whole Pacific.* Lowy Institute. https://policy commons.net/artifacts/2293737/a-security-agreement-between-china-and-solomon-islands-could-impact-stability-in-the-whole-pacific/3054011/

Sørensen, H. (1994). New perspectives on the military profession: The I/O model and esprit de corps reevaluated. *Armed Forces & Society, 20*(4), 599–617. https://doi.org/10.1177/0095327X9402000407

Sprugnoli, G., Rossi, S., Emmendorfer, A., Rossi, A., Liew, S.-L., Tatti, E., di Lorenzo, G., Pascual-Leone, A., & Santarnecchi, E. (2017). Neural correlates of *Eureka* moment. *Intelligence, 62*, 99–118. https://doi.org/10.1016/j.intell.2017.03.004

Stewart, K. (2009). Mission command: Problem bounding or problem solving? *Canadian Military Journal, 9*(4), 50–59.

Stewart, D. (2019). *Developing creative leaders: Understanding innovation within the profession of arms.* Canadian Forces College. www.cfc.forces.gc.ca/259/290/308/286/stewart.pdf

Stofka, J. M. (2010). *Designing the desired state: A process and model for operational design* [Master's thesis]. United States Marine Corps Command and Staff College.

Swallow, A. (1959). The Mavericks. *Critique: Studies in Contemporary Fiction, 2*(3), 74–92. https://doi.org/10.1080/00111619.1959.10689689

Sweller, J. (1988). Cognitive load during problem solving: Effects on learning. *Cognitive Science, 12*(2), 257–285. https://doi.org/10.1207/s15516709cog1202_4

Sweller, J. (2010). Element interactivity and intrinsic, extraneous, and germane cognitive load. *Educational Psychology Review, 22*, 123–138. https://doi.org/10.1007/s10648-010-9128-5

Taura, T., & Nagai, Y. (2017). Creativity in innovation design: The roles of intuition, synthesis, and hypothesis. *International Journal of Design Creativity and Innovation, 5*(3–4), 131–148. https://doi.org/10.1080/21650349.2017.1313132

Taylor, M. Z., & Wilson, S. (2012). Does culture still matter?: The effects of individualism on national innovation rates. *Journal of Business Venturing, 27*(2), 234–247. https://doi.org/10.1016/j.jbusvent.2010.10.001

Teichert, J. (2023). *Boom!: Leadership that breaks barriers, challenges convention, and ignites innovation.* Capital Leadership Books.

Thomas, M. T. (2001). Music for an evolving world. In S. Key & L. Rothe (Eds.), *American mavericks* (pp. 117–119). University of California Press.

Tosey, P., Visser, M., & Saunders, M. N. (2012). The origins and conceptualizations of 'triple-loop' learning: A critical review. *Management Learning, 43*(3), 291–307. https://doi.org/10.1177/1350507611426239.

Tucker, R. B. (2002). *Driving growth through innovation: How leading firms are transforming their futures.* Berrett-Koehler Publishers.

Tukey, J. W. (1977). *Exploratory data analysis* (Vol. 2). Addison-Wesley Publishing.

Turan, H. H., Kahagalage, S. D., Jalalvand, F., & El Sawah, S. (2021). A multi-objective simulation–optimization for a joint problem of strategic facility location, workforce planning, and capacity allocation: A case study in the Royal Australian Navy. *Expert Systems with Applications, 186,* 115751. https://doi.org/10.1016/j.eswa.2021.115751

Valayden, D. (2020). The littoral combat ship, or the designs of liquid sovereignty. In A. S. Moore & S. Pinto (Eds.), *Writing beyond the state: Post-sovereign approaches to human rights in literary studies* (pp. 155–174). Palgrave Macmillan. https://doi.org/10.1007/978-3-030-34456-6_8

Vallerand, A. L., & Masys, A. J. (2022). Shocks and disruptions in defence and security: How to lead by inspiring innovation through ideation? In *Disruption, ideation and innovation for defence and security* (pp. 151–178). Springer.

Vergauwe, J., Willie, B., Hofmans, J., Kaiser, R. B., & De Fruyt, F. (2018). The double-edged sword of leader charisma: Understanding the curvilinear relationship between charismatic personality and leader effectiveness. *Journal of Personality and Social Psychology, 114*(1), 110–130. https://doi.org/10.1037/pspp0000147

Villain, N. (2020). Ultracrepidarianism, cognitive biases, and COVID-19. *Revue de neuropsychologie, 12*(2), 216–217.

von Bertalanffy, L. (1968). General system theory as integrating factor in contemporary science. *Akten des XIV. Internationalen Kongresses für Philosophie, 2,* 335–340. https://doi.org/10.5840/wcp1419682120

Wadham, B., & Connor, J. (2023). Commanding men, governing masculinities: Military institutional abuse and organizational reform in the Australian armed forces. *Gender, Work & Organization, 30*(5), 1533–1551. https://doi.org/10.1111/gwao.12986

Wang, L., Yeung, J. H. Y., & Zhang, M. (2011). The impact of trust and contract on innovation performance: The moderating role of environmental uncertainty. *International Journal of Production Economics, 134*(1), 114–122. https://doi.org/10.1016/j.ijpe.2011.06.006

White, D. J. (1975). *Decision methodology: A formalization of the OR process.* Wiley.

Whitt, J. E. P., & Perazzo, E. A. (2018). The military as a social experiment: Challenging a trope. *Parameters, 48*(2), 5–12. https://doi.org/10.55540/0031-1723.2940

Williams, J. C. (2015). HEART – A proposed method for achieving high reliability in process operation by means of human factors engineering technology. *Safety and Reliability, 35*(3), 5–25. https://doi.org/10.1080/09617353.2015.11691046

Williamson, D. (2023, June 16). The RAAF's frozen middle. *Air/Space Blog.* https://airpower.airforce.gov.au/blog/raafs-frozen-middle

Wrigley, C. (2016). Design innovation catalysts: Education and impact. *She Ji: The Journal of Design, Economics, and Innovation, 2*(2), 148–165.

Wrigley, C. (2017). Principles and practices of a design-led approach to innovation. *International Journal of Design Creativity and Innovation, 5*(3–4), 235–255.

Wrigley, C., & Mosely, G. (2022). *Design thinking pedagogy: Facilitating innovation and impact in tertiary education.* Routledge. https://doi.org/10.4324/9781003006176

Wrigley, C., Mosely, G., & Mosely, M. (2021). Defining military design thinking: An extensive, critical literature review. *She Ji: The Journal of Design,*

Economics, and Innovation, 7(1), 104–143. https://doi.org/10.1016/j. sheji.2020.12.002

Wrigley, C., Nusem, E., & Straker, K. (2020). Implementing design thinking: Understanding organizational conditions. *California Management Review, 62*(2), 125–143. https://doi.org/10.1177/0008125619897606

Wrigley, C., Rana, H., Hinton, P., & Mosely, G. (2020). The Defence by Design framework: Conceptual foundations and potential applications. *Journal of Design, Business & Society, 7*(2), 211–232.

Yigitcanlar, T., Sabatini-Marques, J., Kamruzzaman, M., Camargo, F., Moreira da-Costa, E., Ioppolo, G., & Palandi, F. E. D. (2018). Impact of funding sources on innovation: Evidence from Brazilian software companies. *R&D Management, 48*(4), 460–484. https://doi.org/10.1111/radm.12323

Young, L. (2017). The conservative colonel: How being creative killed your career in the ADF. *Australian Defence Force Journal, 2017*(203), 47–56.

Zisk, K. M. (1993). *Engaging the enemy: Organization theory and Soviet military innovation, 1955–1991.* Princeton University Press. https://doi. org/10.1515/9781400820931

Zweibelson, B. (2015). An awkward tango: Pairing traditional military planning to design and why it currently fails to work. *Journal of Military and Strategic Studies, 16*(1), 11–41.

Zweibelson, B. (2017). Blending postmodernism with military design methodologies: Heresy, subversion, and other myths of organizational change. *Journal of Military and Strategic Studies, 17*(4), 139–164.

Zweibelson, B. (2022, April 15). Understanding emergence: How complexity theory requires getting out of the military's favored Newtonian box. *Medium.* https://benzweibelson.medium.com/understanding-emergence-how-complexity-theory-requires-getting-out-of-the-militarys-favored-bf914e06f1e

Zweibelson, B. (2023). *Understanding the military design movement: War, change and innovation.* Routledge. https://doi.org/10.4324/9781003387763

Zweibelson, B. (2024). Why do militaries stifle new ideas? *Contemporary Issues in Air & Space Power, 2*(1), 1–6. https://doi.org/10.58930/bp38138320

Zwicky, F. (1967). The morphological approach to discovery, invention, research and construction. In F. Zwiky & A. G. Wilson (Eds.), *New methods of thought and procedure* (pp. 273–297). Springer.

Index

Note: Page numbers in *italics* indicate figures, and page numbers in **bold** indicate tables in the text